Disorder and Order in the Solid State

Concepts and Devices

Institute of Amorphous Studies Series

Series editors

David Adler†
Massachusetts Institute of Technology
Cambridge, Massachusetts

and

Brian B. Schwartz
Institute for Amorphous Studies
Bloomfield Hills, Michigan
and Brooklyn College of the City University of New York
Brooklyn, New York

DISORDER AND ORDER IN THE SOLID STATE
Concepts and Devices
Edited by Roger W. Pryor, Brian B. Schwartz, and
Stanford R. Ovshinsky

DISORDERED SEMICONDUCTORS
Edited by Marc A. Kastner, Gordon A. Thomas, and
Stanford R. Ovshinsky

LOCALIZATION AND METAL–INSULATOR TRANSITIONS
Edited by Hellmut Fritzsche and David Adler

PHYSICAL PROPERTIES OF AMORPHOUS MATERIALS
Edited by David Adler, Brian B. Schwartz, and Martin C. Steele

PHYSICS OF DISORDERED MATERIALS
Edited by David Adler, Hellmut Fritzsche, and Stanford R. Ovshinsky

TETRAHEDRALLY-BONDED AMORPHOUS SEMICONDUCTORS
Edited by David Adler and Hellmut Fritzsche

A Continuation Order Plan is available for this series. A continuation order will bring delivery of each new volume immediately upon publication. Volumes are billed only upon actual shipment. For further information please contact the publisher.

Disorder and Order in the Solid State

Concepts and Devices

Edited by

Roger W. Pryor

Energy Conversion Devices, Inc.
Troy, Michigan

Brian B. Schwartz

Institute for Amorphous Studies
Bloomfield Hills, Michigan
and Brooklyn College of the City University of New York
Brooklyn, New York

and

Stanford R. Ovshinsky

Energy Conversion Devices, Inc.
Troy, Michigan

Springer Science+Business Media, LLC

Library of Congress Cataloging in Publication Data

Disorder and order in the solid state.

(Institute for Amorphous Studies series)
Includes bibliographical references and index.
1. Solid state physics. 2. Thin films. 3. Order-disorder models. 4. Amorphous
substances. I. Pryor, Roger W. II. Schwartz, Brian B., 1938–   . III. Ovshinsky,
Stanford R. IV. Series.
QC176.2.D57 1988   530.4′1   88-12574

ISBN 978-1-4612-8299-0   ISBN 978-1-4613-1027-3 (eBook)
DOI 10.1007/978-1-4613-1027-3

© 1988 Springer Science+Business Media New York
Originally published by Plenum Press, New York in 1988
Softcover reprint of the hardcover 1st edition 1988

To
Professor Heinz K. Henisch
with great respect and affection
on the sixty-fifth anniversary of his birth

PREFACE

This Festschrift is an outgrowth of a collection of papers presented as a conference in honor of Professor Heinz K. Henisch on his sixty-fifth birthday held at the Institute for Amorphous Studies, Bloomfield Hills, Michigan. It is our great pleasure to be editors of the Festschrift volume to honor Heinz and his work.

Professor Henisch has a long and distinguished career and has many accomplishments in semiconductor materials and devices. He has made seminal contributions to the understanding of semiconductor switching devices and contact properties. He has an outstanding reputation as an expositor of science. His seminars and lectures are always deep, lucid and witty. He received his doctorate in Physics from the University of Reading and then joined the faculty. In 1963, he accepted a position in the Department of Physics at Pennsylvania State University. While at Penn State, Dr. Henisch broadened his research interest to include the History of Photography. At the present time, Dr. Henisch holds parallel appointments as a Professor of Physics and a Professor of the History of Photography at Pennsylvania State University. He is a Fellow of the American Physical Society, the Institute of Physics, London, the Royal Photographic Society and is a Corresponding Member of the Deutsche Gesellschaft fur Photographie. In addition to his considerable publication in the fields of physics and the history of photography, Dr. Henisch is the founder and editor of the Journal of the History of Photography published quarterly by Taylor and Francis, Ltd., London.

The papers in this Festschrift reflect only a small portion of the breadth of the total scientific contribution of Professor Henisch, his students and his associates. The papers are a selection of "snapshots" of the activities and influences of Heinz Henisch. This volume contains 17 papers by 30 authors from 6 countries around the world. We are grateful to the many friends and associates of Dr. Henisch for their contributions to this volume.[*]

---

[*] A number of excellent papers related to the history of photography were also presented at the Henisch Festschrift conference at the Institute for Amorphous Studies. Those papers are published as a separate volume and edited by Dr. Kathleen Collins.

The topics discussed in this volume range from superconductivity theory to experimental measurements of piezoresistivity, and from the long-range disorder of the amorphous state to various types of short-range order and disorder. The largest number of papers, 11, address the measurement and the modeling of different aspects of the electronic properties of solid-state devices and thin solid films, both single and multilayer. Other papers consider the various effects of order/disorder on the properties of solids and the fabrication of amorphous films and fibers.

The editors wish to express their gratitude to Ghazaleh Koefod for her many contributions, without which this volume would not have been completed.

Roger W. Pryor

Brian B. Schwartz

Stanford R. Ovshinsky

CONTENTS

INTRODUCTION

The HENISCH FESTSCHRIFT is a collection of papers presented in honor of the sixty-fifth birthday of Professor Heinz K. Henisch. His contributions to the understanding of semiconductors, their contacts and to the science of electronic materials has been extensive. His book on "Rectifying Semiconductor Contacts" has long been considered the definitive document in that field. The papers in this collection only partially reflect the extent of the total scientific contribution that he, his students and his associates have made. These papers are presented as a brief summary of the current activities of a great scientist and his associates.

This volume is international in scope and contains 17 papers by 30 authors from 6 different countries around the world. The topics covered in this volume range from superconductivity to piezoresistivity and include many different kinds of states of disorder such as the amorphous states and the state with long-range order but with no short-range order. The largest number of papers, 11, on amorphous thin film concepts and devices, address the measurement and modeling of different aspects of the electronic properties of solid state devices and thin solid films, both single and multilayer. The second largest number of papers, 4, consider the various effects of order/disorder on the properties of solids. The topics discussed include superconductivity, piezoresistivity, magnetism, and the various types of short-range order and disorder. The last category, on films and fibers, contains 2 papers that treat the fabrication of amorphous films and fibers.

The initial section in this volume is titled "AMORPHOUS THIN-FILM CONCEPTS & DEVICES." The first two papers in this section concern the conductivity of thin amorphous films. The first paper is by Sir Nevill Mott of the University of Cambridge on the "Mobility of Electrons in Non-crystalline Materials." Sir Nevill discusses the validity of the polaron hypothesis as it relates to the question of the fundamental conduction mechanism in amorphous semiconductors. He calculates an expression for the mobility based on the inelastic diffusion length and the time between inelastic collisions with phonons. He obtains a value of the mobility for amorphous silicon that agrees well with current experimentally derived values. Sir Nevill concludes by suggesting areas of research that are in need of further experimental exploration. The next

1

paper is by H. Fritzsche, H. Ugur, J. Takada, and S. H. Yang of the University of Chicago on "Metastable Nonlinear Conductance Phenomena in Amorphous Semiconductor Multilayers". In this paper, Fritzsche et. al. present a series of experimental observations on amorphous semiconductor multilayer films and the conclusions that can be made from their measurements. The films used were of two types: compositionally modulated multilayers (a-Si:H-a-SiN:H) and doping-modulated multilayers (a-Si npnp). The compositionally modulated multilayer films exhibit no unusual behavior at low applied biases (<1 volt). At high applied bias (300 volts), excess dark conductance was observed after brief photoillumination. The excess dark conductance can be removed by photoillumination of the sample at no bias or by heating. The doping-modulated multilayer films also show field related conductance anomalies. This paper shows that the persistent photoconductivity (PPC) effect is reduced by applied bias during light exposure. It is concluded that there may be a link between these effects and the glass-like behavior of the bonded hydrogen network in a-Si:H films that mediates the equilibration of defects and dopants.

The next seven papers in the "AMORPHOUS THIN FILM CONCEPTS & DEVICES" section are on device technology, device behavior and device theory. The first paper is by Roger W. Pryor of Energy Conversion Devices entitled "Recent Developments in Ovonic Threshold Switching Device Technology." This paper discusses the development of two new device designs and processing techniques for the fabrication of thin film, high current, DC stable, chalcogenide glass threshold switches and presents the first results showing hydrogen as an electronically active material modifying the behavior of the chalcogenide switching glasses. It is shown that hydrogen significantly improves the threshold voltage stability and reproducibility. Experimental data are presented on the current density during the ON-State, which confirms the principle of the expanding filament.

The next device paper is by D. N. Bose of the Indian Institute of Technology. His paper discusses the use of chalcogenide glasses in "Amorphous Chalcogenide Microwave Switches." He shows that thin chalcogenide films can act as microwave switches in the X-band. Measurements were made on both threshold and memory glass compositions using the switches to modulate an applied X-band signal. A mechanism for the change in conductance is presented.

Melvin P. Shaw of Wayne State University presents a paper on "Switching and Memory Effects in Thin Amorphous Chalcogenide Films - Thermophonic Studies." In this paper, the author briefly reviews the experimentally observed characteristics of threshold switching devices before presenting his additional observations by electroacoustic spectroscopy (thermophonics). Data taken simultaneously on the voltage, current and acoustic signals demonstrate that electronic switching occurs well before the thermally generated acoustic signals under normal conditions. Data

2

and analysis are also presented on devices operated under other than normal conditions.

The role of relaxation semiconductors is considered in next two papers. "Interpretation of Recent ON-State and Previous Negative Capacitance Data in Threshold Chalcogenide Amorphous Switch" is the title of the first such paper by G. C. Vezzoli of U. S. Army Materials Technology Laboratory and M. A. Shoga of Hughes. These authors show that the blocked on-state requires approximately 65 ns to form in thin chalcogenide threshold switching devices. For times less than the critical time of formation, the I/V curve of the switch appears to be metallic in nature. It is argued that these results support a recombination single injection model for the mechanism of threshold switching. In the paper "Carrier Injection into Low Lifetime (Relaxation) Semi-Conductors" by J. C. Manifacier, Y. Moreau, and R. Ardebili, the mechanism of conduction in relaxation semiconductors is the main theme. The authors review the equations governing the behavior of electrons and holes in the general case for semiconducting materials. They proceed to do a comparison between an analytical and a numerical solution to these equations for both the lifetime and relaxation cases for two types of junctions. The calculations presented agree well with previously published experimental data.

Peter T. Landsberg of the University of Florida and the University of Southampton, U.K. contributed a paper on "A Semiconductor Model for Electronic Threshold Switching." The author presents a model for a method of electronic switching based on electron and hole concentration using the stability criterion of the recombination, generation parameters in impact ionization space to explain the occurrence of switching. In this model, a switching transition occurs as a result of a field-driven boundary crossing from one region of stability to a region in which different stability criteria apply.

In the next paper, "Proper Capacitance Modeling for Devices with Distributed Space Charge" by Roberto C. Callarotti of INTEVEP S. A., a detailed method of modeling device capacitance is presented. The primary feature of this model is that the bulk of the device is divided into slices and each slice is electrically modeled as a parallel resistor and capacitor. All the electrical analogs are then connected in series for calculation. An example is given using a Metal-Insulator-Metal device.

P. E. Schmidt of IVIC and R. C. Callarotti contributed a paper on "On the Impedance Calculation of Thick MIM Barriers". This paper demonstrates methods for the calculation of the incremental AC characteristics of thick metal-insulator-metal barriers. A generalized approach results in the formulation of coupled ordinary differential equations in terms of a complex AC carrier density. This simplified approach utilizes an incremental circuit representation as a reasonable approximation. A low frequency model is presented and discussed in detail.

K. B. R. Varma, K. J. Rao and C. N. R. Rao present the

final paper in the "AMORPHOUS THIN FILM CONCEPTS & DEVICES" section entitled "Dielectric Behavior of Amorphous Thin Films." They present an analysis of the dielectric behavior and the ultramicrostructures of RF sputtered amorphous and recrystallized high dielectric constant films. The behavior of these films are analyzed using a cluster model. It is concluded that amorphous films of ionic dielectrics exhibit large dielectric constants and ferroelectric-like dielectric anomalies in the region of crystallization. It is also concluded that the dielectric anomalies in the crystallized state and in the amorphous films are not directly related.

"ORDER & DISORDER" is the next section of this volume and includes those papers that analyze the influence of the relative degree of order or disorder on some of the observable properties of solids. The topics that are included within this section are superconductivity, various kinds of short- and long-range order and disorder, piezoresistivity and magnetic domain patterns.

The first paper in this section is by S. R. Ovshinsky (inventor of the threshold switch) of Energy Conversion Devices entitled "A Personal Adventure in Stereochemistry, Local Order, and Defects: Models for Room Temperature Superconductivity." The author presents a new and unique approach to the theory of superconductivity. In his discussion, he covers primarily two topics. The possibility of room temperature superconductivity being the mechanism of conduction in the Ovonic Threshold Switch (OTS) during the ON-State and the establishment of high-temperature superconductivity by a simultaneous, two-atom valence transformation process. In the case of the OTS, the formation of the superconductive filament occurs in the solid state plasma. Bose particles are formed when the carriers can pair as a result of the negative correlation energy in the lone pair configurations. It is proposed that "such pairs should exhibit a Bose condensation well above room temperature." In the new ceramic oxide high-temperature superconductors, a different mechanism is proposed. The basis of that mechanism is the establishment of a simultaneous, two-atom valence transformation. This is accomplished by the interaction of copper "sheets" and the copper "chains" through the oxygen "pump." This results in "superexchange coupled spin pairs on the chains and sheets" that "are mobile, strongly bound, spinless composite particles which obey Bose statistics."

A. H. Majid, W. F. Anderson and R. L. Osgood of the Pennsylvania State University and T. Madjid of the Department of the Air Force present the next paper entitled "Phenomenology of Antiamorphous Order." This paper treats the phenomenology of metal/insulator/metal superlattice structures, where the layers are sufficiently thin to be aggregates. Such aggregates are an experimental approach to the formation of a state in which the material has long-range order without having short-range order. These multilayer aggregate structures exhibit a broad range of interesting phenomena, including switching. The methods of fabrication and the observed results are discussed.

The "Piezoresistivity in Semiconducting Ferroelectrics" paper by Ahmed Amin of Texas Instruments reviews the piezoresistive effect in semiconducting polycrystalline barium titanate and its solid solutions with lead and strontium titanate under different elastic and thermal boundary conditions. An analysis of this phenomenology is presented in terms of current ferroelectric and grain boundary potential models. The results are compared to that for silicon and germanium.

The final paper in the section is by Walter M. Fairbairn of the University of Lancaster, U. K. on "Domain Patterns in Helical Magnets". The author presents an analysis of helical domains in rare earth materials. He discusses the effects of static, time-varying, and non-uniform applied fields in addition to temperature related effects and concludes that helically-ordered domain structures are very stable and not easily altered.

The last section in this volume is on "FILMS & FIBERS" and contains two papers. The first paper is on "Silicon Nitride Films Formed with DC-Magnetron Reactive Sputtering" by Napo Formigoni of Energy Conversion Devices. The author discusses the development of both the hardware and the process to make silicon nitride films using a DC-magnetron sputtering system with a silicon target in a reactive nitrogen/argon gas environment. Detailed information is presented on the design of the equipment and the quality and uniformity of the deposited films. These films are used as the passivation layer on the new thin film Ovonic Threshold Switching devices.

The final paper is "Optical Fibers for Infrared from Vitreous Ge-Sn-Se" by L. Haruvi, J. Dror, D. Mendleovic and N. Croitoru of Tel-Aviv University. The authors discuss the experimental methods and data obtained during the fabrication of chalcogenide fibers for use at a wavelength of 10.6 microns. Several compositions were formed into glass. The glass drawn into fibers showed an attenuation of 6 db/cm, as compared to 10 db/cm for GeSe(4).

The three sections of this volume represent an important collection of new results and analyses in a broad cross-section of solid state science and technology. These papers honor Professor Henisch for his many contributions to science and technology. It is hoped that this Festschrift will contribute to the stimulation of further work toward the understanding of the phenomenology of solids, devices and contacts.

Roger W. Pryor

Brian B. Schwartz

Stanford R. Ovshinsky

HEINZ HENISCH FESTSCHRIFT

MOBILITY OF ELECTRONS IN NON-CRYSTALLINE MATERIALS

Nevill F. Mott

Cavendish Laboratory
University of Cambridge
Cambridge CB3 OHE, U.K.

INTRODUCTION

On the occasion of Heinz Henisch's retirement I remember him for many things. His early book, published by the Oxford University Press, on rectifying semiconductor contacts, written I think when Heinz was still at Reading University in the U.K.; his creation and editorship of the hugely enjoyable journal, "The History of Photography"; but above all for his experimental work on the Ovonic threshold switch, which I followed with close interest while trying to make a theory of what was happening in this device (Mott 1969, 1971, 1975).

Finally we brought together the work of Heinz and of Dave Adler (of whose sad death I learned recently) and my theoretical speculations in a review article (Adler, Henisch & Mott 1978). We were far from getting a complete answer, particularly about the switching process itself. But we did obtain a consistent, if speculative, model of the on-state of the switch. In the last ten years I have not seen further work on the basic mechanism; I would like, then, to look back on this paper and to see how it stands up to more recent work on the conduction process in non-crystalline materials.

THEORY

In the conducting channel in the layer of chalcogenide glass electrons and holes are injected from the electrodes, forming a plasma. Both the density of electrons and the current density are independent of the total current; it is the cross-sectional area that varies. The thickness of the channel is normally greater than the thickness of the film. Some dynamical process is responsible for the voltage across the contacts which allows injection. The channel is not hot, and its conductivity is not due to temperature. It is also suggested that the density of carriers, in the range 10^{18}-10^{19} cm^{-3}, might be that at which the density of the electron-hole gas is a minimum, though this is perhaps doubtful; the existence of a minimum depends on the electron and hole gases being degenerate, and this is unlikely at the temperature slightly above room temperature in the conducting channel, unless the density of states is rather low; however as we shall see below, this is possible.

I want particularly to discuss the high mobility of the carriers, shown by the experiments described to be about 10 cm^2/V sec. This has always seemed to me the strongest evidence that the carriers are not polarons. Emin, Seeger and Quinn (1972) give evidence for an activated mobility and seek to explain it by postulating the existence of polarons; Mott and Davis (1979) give arguments to show that it can be explained, as is usual in hydrogenated amorphous silicon (a-Si-H), through the presence of a mobility edge. These high mobilities seem hardly consistent with polaron formation. On the other hand the high density of electrons in the channel could ensure that the Fermi energy lies above the mobility edge E_c. In a-Si-H extensive recent research shows that the density of states at the mobility edge is given by

$$N(E_c) \sim 2 \times 10^{21} \text{ cm}^{-3} \text{ eV}^{-1}$$

and that N(E) drops linearly towards zero over a range of E of perhaps 0.05 eV, below which the small values characteristic of the exponential tail begin. There are thus $\sim 10^{20}$ cm^{-3} states in the conduction band below the mobility edge. This differs by 10 from the highest estimate of electrons and holes in the switch, but we do not know the density of states near E_c in the chalcogenide glasses; it could be lower than in a-Si-H.

We do - as we did not in 1978 - know how to calculate the mobility just above E_c. This follows from the scaling theory of Abrahams et al. (1979). If we write the conductivity

$$\sigma = \sigma_0 \exp\{-(E_c-E_F)/kT\} \ ,$$

where E_F is the Fermi energy, then we believe (Mott 1987) that

$$\sigma_0 \simeq 0.03 \ e^2/\hbar L_i \ ,$$

where L_i is the <u>inelastic</u> diffusion length for an electron with energy just above E_c. L_i is given by

$$L_i = (D\tau_i)^{1/2}$$

where τ_i is the time between (inelastic) collisions with phonons and D the diffusion coefficient resulting from elastic scattering by disorder. We find D from the equation

$$\sigma_0 = e^2 N(E) \ D \ ,$$

giving

$$\sigma_0 = (0.03)^{2/3} \ e^2 (N/\hbar^2 \tau_i)^{1/3}.$$

The value of μ at the mobility edge is given by

$$\sigma_0 = eN(E_c)kT\mu \ ,$$

so

$$\mu = \{e(0.03)^{2/3}/kT\} \ (1/\hbar\tau_i N(E_c)^2)^{1/3} \ .$$

In a-Si-H reasonable values of these quantities lead to $\mu \sim 10 \text{ cm}^2/\text{V sec}$, in agreement with experiment; a smaller value of $N(E_c)$ which we have suggested might be the case for a chalcogenide glass would increase τ_i, so the dependence on the density of states will not be great. We have no experimental value of τ_i, but since it appears only to the power $^1/_3$ we should not expect any large difference from the value of μ for a-Si-H, which is what Adler et al. deduce from the experimental work of both Henisch and Adler.

Experimental values of the quantities concerned would, we believe, be of considerable interest in elucidating the conduction process in those materials, particularly in the switch.

REFERENCES

Adler D., Henisch H.K. and Mott N.F., Rev. Mod. Phys. 50:209 (1978).

Abrahams E., Anderson P.W., Liccardello D.C. and Ramakrishnan T.W., Phys. Rev. Lett. 42:693 (1979).

Emin D., Seeger C.H. and Quinn R.K., Phys. Rev. Lett. 28:813 (1972).

Mott N.F., Contemp. Phys. 10:129 (1969); Phil. Mag. 24:91 (1971); ibid 32:159 (1975); Phil. Mag. B (in press).

Mott N.F. and Davis E.A., "Electronic Processes in Non-crystalline Materials", 2nd edn., Oxford University Press, Oxford (1979).

METASTABLE NONLINEAR CONDUCTANCE PHENOMENA IN

AMORPHOUS SEMICONDUCTOR MULTILAYERS

H. Fritzsche, H. Ugur*, J. Takada†, and S.-H. Yang††

James Franck Institute
The University of Chicago
Chicago, IL 60637

Nonlinear current-voltage characteristics in amorphous semiconductor multilayers are reviewed with the hope of gaining a better understanding of conductance anomalies in heterogeneous semiconductors. Electronic transport parallel and perpendicular to the layers of npnp doping-modulated multilayers as well as of $a-Si:H/a-SiN_x:H$ multilayers are considered. The npnp multilayers can be brought into different metastable states by cooling from an equilibration temperature $T > T_e$ with $T_e = 130^o C$ while a bias is applied. These metastable states exhibit greatly different nonlinear current-voltage characteristics.

INTRODUCTION

The difficulty of studying the electronic transport properties of semiconductors is often not fully appreciated. We are not referring to the complicated task of unravelling the interactions of the charge carriers with phonons, defects, and impurities, but to the problem of understanding the boundary conditions which may affect the experimental results, in particular the contacts. The bulk transport properties are not easily extracted from the experimental data unless the boundary conditions of the contacts are known and this in turn is possible only after a detailed understanding of the given semiconductor is available. H.K. Henisch has made immense contributions by pointing the way towards a solution or at least elucidating the net of problems that arise when one disregards the influence of contacts on transport measurements.[1,2]

It would be futile for us to try to add something of significance to this topic which has been a major research area of H.K. Henisch. Instead, we wish to discuss some other factors which can greatly affect transport measurements and which prevent an easy interpretation in terms of bulk properties of a semiconducting material. These are compositional and structural heterogeneities. A number of semiconducting materials exhibit a variety of anomalies in their electronic properties[3] such as nonohmicity, very long time constants for establishing a constant current after a change in applied voltage, anomalously large current fluctuations, and an excess conductance that persists for hours or days after illumination, called persistent photoconductivity. The detailed origin of these phenomena remains unclear, but features common to these materials are a heterogeneous structure and the presence of sizable internal electric fields.

---

*Present address: Research Institute Basic Sciences, Gebze, Kocaeli, Turkey.

†Present address: Central Research Lab. Kanegafuchi Chemical Industry Co.,
                   Kobe 652, Japan.

††Present adress: Department of Physics Education,
                   Kyungpook National University, Daegu 635, Korea.

A good way to investigate these observations is to prepare samples with well controlled heterogeneities with a relatively simple geometry. We chose two kinds of amorphous semiconducting multilayers. The first, termed doping-modulated multilayers,[4-7] are made of ultrathin alternating layers of p-type and n-type doped amorphous silicon. The second are called compositionally modulated multilayers[8-11] and are made of alternating layers of amorphous silicon and silicon nitride. The essential difference between the two kinds of multilayers is that in the doping-modulated materials the band gap is constant but the internal potential is spatially modulated, whereas in the compositionally modulated-materials the dominant modulation parameter is the magnitude of the band gap.

We shall review here some unusual electrical properties of these two kinds of amorphous multilayers. They have much in common with ordinary heterogeneous semiconductors which also contain these modulation parameters but in a random fashion. In contrast to crystalline superlattices, amorphous multilayers can be made quite simply because there is no requirement to match lattices at an interface. The ones considered here were prepared by rf plasma assisted chemical vapor deposition of SiH_4 containing PH_3 or B_2H_6 for n-type or p-type doping,[4,5] or NH_3 for the silicon nitride layers.[8,9] The amorphous silicon prepared in this manner on a glass substrate held at $250^o C$ as well as the nitride layers contain about 8 at. % hydrogen. We therefore refer to these materials as $a-$Si:H and $a-SiN_x:H$, respectively.

COMPOSITIONALLY MODULATED MULTILAYERS

Observations

Let us consider the coplanar conductance of multilayers[11] that consist of 32 layers of 260Å thick $a-$Si:H and 33 layers of 40Å thick $a-SiN_x:H$. The total thickness of each film is close to 1μm. Electrical contacts to all layers were made by scratching the films with a diamond scribe and then depositing either carbon paint or evaporated NiCr electrodes. The electrodes were 1 cm wide and 0.2 cm apart.

Fig. 1 shows a summary of the observed anomalies of the coplanar conductance of these structures.[11] At a low applied bias of 1V we find no surprises, see dashed curve. When such a small bias is applied at t=0, a steady current appears without delay. A photoconductance marked $G_P(1)$ appears when the sample is exposed to light at t=60 min and disappears when the light is turned off. However, the behavior is very different when a 300V bias is applied at t=0, see full curve. It takes about one

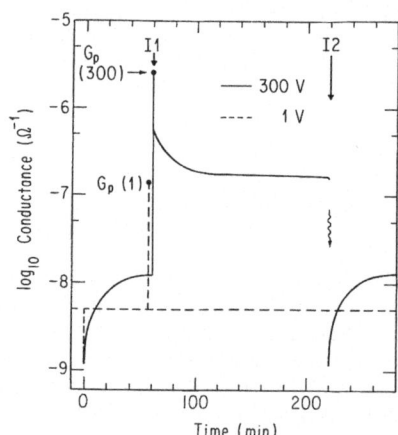

Fig.1   Time dependence of the coplanar conductance of a $a-$Si:H/$a-SiN_x:H$ multilayer film as bias is applied at t=0. 300V bias, solid curve; 1V bias, dashed curve. At t=60 min illumination for 7 sec. with bias (I1). At t=220 min illumination for 7 sec at zero bias (I2), followed again by a bias. After Ref. 11.

hour before the conductance G reaches a steady state value. This value is larger than the low bias G. The brief illumination at t=60 min produces a photoconductance $G_P(300)$ which greatly exceeds $G_P(1)$, but even more surprising is the fact that a metastable excess conductance state is maintained after the light is turned off. Even 3h after the brief light exposure G is still a factor 10 larger than the initial G at 300V. Removal and reapplication of the 300V bias does not change this excess conductance state. At t=220 min the bias is removed and the sample is exposed to light for a few seconds (I2 in Fig. 1). This procedure restores the original state. Evidence for this is that reapplication of the 300V bias again yields the original slow rise of G lasting about 60 min. This experiment shows that the anomalous effects occur only at biases that exceed about 50V. These large biases correspond to quite low fields of $2\times10^2 - 2\times10^3$ V/cm assuming the potential drop is uniform. These fields are very small compared to those that are normally associated with nonlinear phenomena in single layer films of $a-$Si:H. The decay of the light-induced excess conductance at larger voltages is extremely slow at room temperature and below but proceeds more and more rapidly at higher temperatures. The excess conductance created at $24^o C$ disappears upon annealing near $T_A=160^o C$. T_A is lower when the light-induced excess conductance is created at a lower temperature. Fig. 2 shows for instance the creation of the excess conductance at 130 K and 170 K, respectively. The metastable states produced in this manner anneal away near room temperature.

We may summarize these observations as follows. No anomalies are observed at small biases, but for biases exceeding 20V or average fields larger than 100 V/cm we find:

(i)   a superlinear I-V characteristics,

(ii)   an equilibration time that grows with increasing bias,

(iii)   a large metastable excess conductance that is established by a brief light exposure in the presence of a large bias,

(iv)   removal of the metastable excess conductance by a brief light exposure while no or only a small bias is applied, and

(v)   annealing of the excess conductance at temperatures exceeding (by about $150^o C$) the one at which the excess conductance was established.

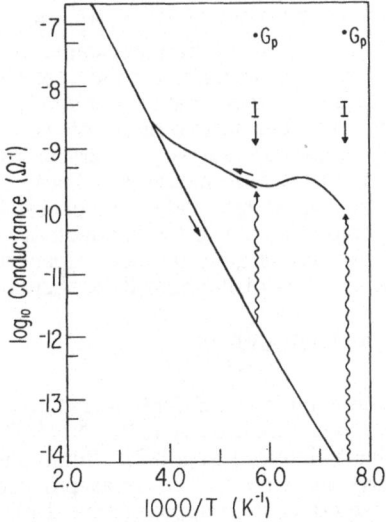

Fig.2   Temperature dependence of coplanar dark conductance before and after the creation of photoinduced excess conductance with 300V bias at low temperatures. Photoconductance values are shown as G_P. Sample is the same as in figure 1. After ref. 11.

These observations are neither caused by the unusual thinness of the semiconducting layers nor by some deep trapping of charge carriers at the interfaces between the insulating and semiconducting layers. We convinced ourselves of that by measuring one thin layer of amorphous silicon sandwiched between two layers of silicon nitride and did not find any of these conductance anomalies.[11]

The anomalies first appear when two layers of a−Si:H are separated by a silicon nitride insulator forming a capacitor structure. Essentially no new phenomena occur when more capacitor structures are stacked on top of one another forming a multilayer. A further requirement for finding the anomalies is that the semiconductor be thin compared to the screening length. This suggests the following explanation.[11] As soon as there are inhomogeneities in the semiconducting layers or nonuniformities in their thickness, the applied bias will cause in a given a−Si:H layer a potential drop that is not uniform and whose spatial profile will in general differ from that of the adjacent a−Si:H layer. As a consequence there will be potential differences across the insulating layers leading to space charges. These bias-induced space charges increase with the applied bias and change the local conductance. This in turn will change the potential profile of the applied bias and the space charges. When the semiconducting layers are thin compared to the screening length, the space charge regions will extend throughout the layer thickness. The total conductance of any layer will then be determined by the topology of these induced space charge regions. Unless a depletion region extends across the whole width of the sample, there will always be a percolation path of higher conductance between the electrodes. Even if the conductance of one layer is diminished, its contribution to the total conductance will be less than the contribution of layers with enhanced conductance. Furthermore our use of intrinsic a−Si:H makes it likely that space charges produce conducting electron as well as conducting hole accumulation regions instead of depletion regions.

Besides inhomogeneities in the a−Si:H layers, uneven and unequal contact resistances can cause local potential differences between layers and hence charged regions on the nitride capacitors. This probably happened in our samples because the experimental results are changed quantitatively even though not qualitatively when the contacts are removed and reapplied.

The long time to establish a constant current after changing the magnitude of the applied bias is in this picture caused by the relatively large capacitance across the nitride layers combined with the high sheet resistance of the thin a−Si:H layers. These RC time constants are of the order of $10^3 - 10^4$ sec near room temperature.

Light has two effects. It reduces the time constants to less than a second because of the high photoconductivity. This accounts for the observations (iii) and (iv), that a brief illumination establishes and removes the excess conductance while a high and a low bias is applied, respectively. The second effect of light is that it will increase the conductance inhomogeneity unless the intensity is very uniform and the absorption depth is large compared to the total film thickness. These conditions were not met in our experiment. Observation (v), the removal of the excess conductance by annealing follows from this model because the rise of the conductivity with temperature rapidly shortens the time constants for discharging the local capacitors. This capacitor model appears to explain the main features of the high field phenomena in these multilayers.

DOPING-MODULATED MULTILAYERS

The most striking observation in npnp doping-modulated multilayers made of a−Si:H is that of persistent photoconductivity (PPC).[4,5,12−16] This is a metastable excess conductance state that is induced by a brief illumination and then decays exceedingly slowly over days and weeks. The decay proceeds the faster the higher the temperature which causes the effect to disappear above $150°C$. The detailed processes leading to the PPC effect are not understood. Even though this is a fascinating problem we will not discuss it in detail here because it is not a phenomenon which is only observed at large applied biases. Nevertheless there seems to be a connection between PPC and the field-induced conductance anomalies which will be described in the following. For one, both are observed in doping-modulated films having n-type and p-type layer thickness between 150 and 1000Å. We believe this is so because the internal field modulation vanishes when the layers become too thin. On the other hand, when the layers are too thick, a large part of the layers becomes field free.

The bias-induced anomalies discussed here have been observed[6,7] on roughly 1μm thick samples consisting of alternating n-layers and p-layers, each between 250 and 500Å thick, and deposited by rf plasma decomposition of silane with gas doping levels of 100 ppm PH_3 or B_2H_6, respectively. We shall discuss both coplanar and transverse conductance measurements. For the latter, the samples were deposited at 250^oC on conducting and transparent In-doped SnO covered glass and later contacted on top with 0.25 cm diameter NiCr dots. The activation energy of the coplanar conductivity was E_a=0.51 eV and the prefactor σ_o=300$\Omega^{-1}cm^{-1}$. The corresponding values of the transverse conductance G measured at 0.1 V are E_a=0.91 eV and G_o=7×10$^4\Omega^{-1}$ for samples having 500Å thick layers.

Coplanar Conductance

The following will only be a phenomenological description of our experimental results because we cannot yet explain them.

Fig. 3 shows the coplanar conductance of a d_n=d_p=500Å sample measured at 24^oC with decreasing applied bias.[6] The measurements are independent of the bias polarity. One observes two conductance states labelled high G and low G. They coincide in the ohmic regime below 5V (across an electrode separation d= 0.27 cm). The high G curve is obtained whenever the sample has been allowed to rest at 24^oC or after it has been annealed, both without an applied bias. The low G curve on the other hand is obtained when the sample is cooled from an annealing temperature of 170^oC while a bias of 300V was maintained. This bias-cooling produces some changes in the material or in the potential and space-charge distribution which yields the low conductance state. It is surprising that these changes only affect the high voltage and not the low voltage part of G(V). As already mentioned above, these changes produced by bias-cooling relax and disappear when the sample is allowed to rest at zero or small biases. Since the rested state yields the high G curve of Fig. 3 we can measure the time dependence of this relaxation by resting a bias-cooled sample and applying only briefly a measuring voltage V_m=140V in order to monitor the recovery from the low G to the high G state. At room temperature this takes about one hour as shown in Fig. 4. On the other hand, the low G state can be maintained without a noticeable change at room temperature by not removing the cooling bias, in this case V_c=300V. Hence

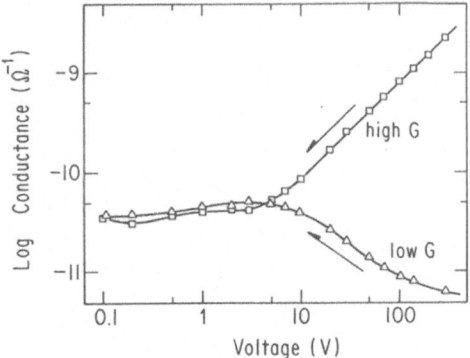

Fig.3   Bias dependence of the coplanar conductance of doping-modulated a−Si:H multilayers at 24^oC after cooling without bias (high G state) and after cooling with 300V bias applied (low G state). After Ref. 6.

15

the low G state is stable at large bias. In contrast, the high G state is not stable at large bias but the relaxation is very slow. For instance after 70h with V_m=140V applied at $24^o C$, G typically decreases by a factor 5. About 10 min after applying the high bias to a rested sample the current decay is so slow that one can measure the high G curve shown in Fig. 3, or one can study the thermal annealing of the high G state.

The thermal annealing is shown by the V_c=0 curve in Fig. 5. A zero cooling-bias V_c=0 yields by definition the high G state. We chose V_c=300V to demonstrate in Fig. 3 the low G state. It should be obvious that bias-cooling with other values of V_c will produce metastable conditions in the multilayer sample that lie in between or, if we choose V_c>300V, conditions that lie below the low G state shown in Fig. 3. The annealing curves of several other conditions obtained by cooling with other biases V_c are shown in Fig. 5. Above $130^o C$ the differences disappear and all G(V) curves resemble the low G state.

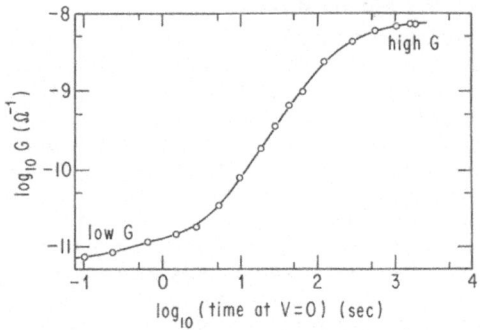

Fig.4   Conversion of the low G to the high G state at $24^o C$ measured with 200V bias as a function of accumulated resting time at zero bias. After Ref. 6.

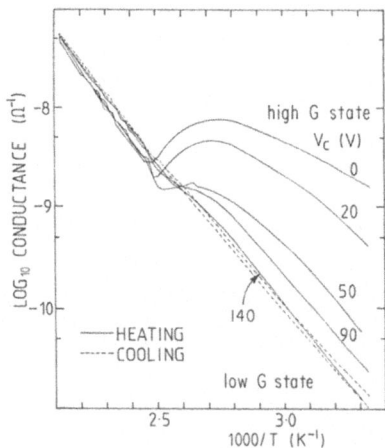

Fig.5   Conductance measured with voltage V_m=140V while heating and subsequent cooling the sample. The different curves correspond to bias-cooled states obtained by cooling the sample with V_c applied. V_c=0 is called the high G state, while larger values of V_c yield lower G values. After Ref. 6.

Without understanding the details one might describe the observations in the following way. An equilibrium condition exists for each applied bias. Equilibrium is rapidly reached above $T_e=130°C$. Bias-cooling below T_e preserves this condition. Changes from one to another equilibrium bias-state, demanded by changing the bias, occur very slowly below T_e. A change from the $V_c=300$ V state to the $V_c=0$ state takes about 1 hour at room temperature according to Fig. 4. A change from the $V_c=0$ to a high V_c state (150 - 300 V) takes a much longer time.

The latter equilibration time is much too long to be caused by space charge relaxations as in the compositionally-modulated multilayers. This problem is very similar to the very slow decay of the persistent photoconductance. This similarity is shown in Fig. 6 where we compare the temperature dependence of the decay time τ of (a) the high G state and (b) the PPC state. Since the decay is not governed by a single decay time but instead by a stretched exponential decay,[17] we plot in Fig. 6 various times τ_x which are defined as the time during which the current drops to half the value it had x minutes after applying the high measuring voltage V_m to the high G state or, in the case of PPC, after stopping the light exposure.

The equilibration temperature $T_e=130°C$ observed in Fig. 5 has the same value as the temperature below which the equilibration of the donor and defect concentrations in n-type $a-$Si:H become frozen-in.[18] Moreover, it was demonstrated by bias-annealing that the equilibrium concentrations can be altered by shifting the Fermi level E_F above T_e before quenching.[19] Such a metastable state produced by bias-annealing and subsequent quenching has been found to relax exceedingly slowly at room temperature similar to the slow decay of our high G state at large biases V_m.

If we are dealing here with the same donor-defect equilibration process then one would expect different results when the sample is either cooled at different speeds or cooled rapidly from different anneal or quench temperatures.[18] The results of these two experiments[6] are shown in Fig. 7 and 8, respectively. The quenching was carried out with $V_c=0$. Essentially the same high G state, measured with $V_m=140$V, was obtained in each case. We have not yet compared the bias-cooled state $V_c=300$V for different quench rates and quench temperatures. The possible relation between the defect equilibration and our bias-induced conductance anomalies requires more detailed exploration.

There is, however, a relation between PPC and the bias effects discussed here besides the one suggested earlier by the similar magnitudes of the relaxation times shown in Fig. 6. We measured the PPC conductance 10 min after exposing the sample for 15s to 32 mW/cm^2 heat-filtered white light while a coplanar bias was applied.[6] The PPC conductance was measured in the ohmic regime with $V_m=1$V. Fig. 9 shows that the PPC conductance is decreased by a factor 4 as the applied bias during illumination is increased to 130V. The decrease starts at about the same voltage at which the nonlinear effects in G begin (see Fig. 3).

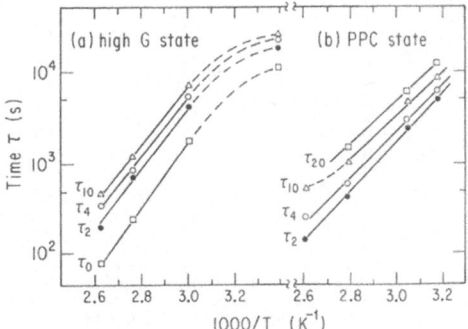

Fig.6   Decay times τ as a function of temperature for (a) the high G state and (b) the PPC state. τ_x is the time for the current to drop to half its value after waiting for x minutes. After Ref. 6.

In the Introduction we mentioned that many heterogeneous semiconductors exhibit in certain bias regimes anomalously large current fluctuations. These are also observed in our samples particularly near the equilibration temperature T_e; see for instance Figs. 5, 7 and 8.

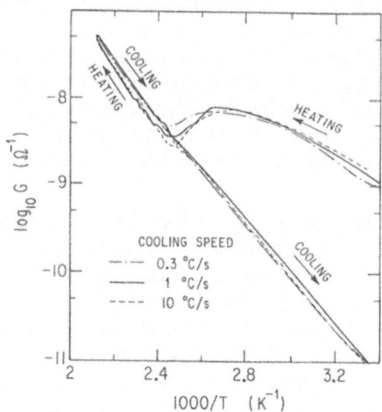

Fig.7　Annealing of the high G state measured with V_m=140V after quenching the sample without bias from $200°C$ with various cooling speeds. d=0.27 cm. After Ref. 6.

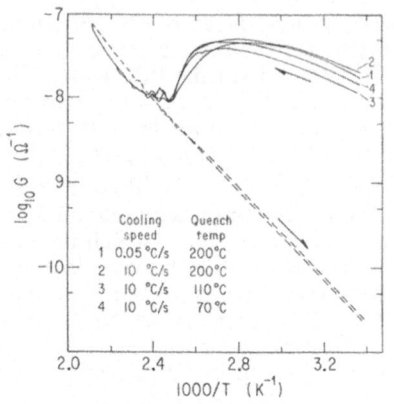

Fig.8　Annealing of the high G state measured wih V_m=140V across an electrode separation of d=0.1 cm after quenching from different temperatures. After Ref. 6.

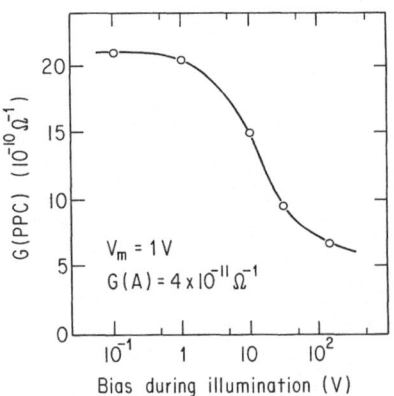

Fig.9

Effect of bias applied during light excitation on the light-induced excess conductance (PPC) measured with V_m=1V. After Ref. 6.

Our observations of the coplanar conductance anomalies can be summarized as follows:

(i) Different states or conditions can be obtained by cooling a sample from $T > T_e$ while different bias values V_c are applied. $T_e \sim 130^o C$.

(ii) The bias-cooled states have the same conductance at small voltages but different G(V) at large voltages.

(iii) The bias-cooled states relax very slowly below T_e but anneal and equilibrate above T_e. The annealing curves closely resemble that of the PPC state.

(iv) The bias-cooled states appear to depend neither on the cooling rate nor on the anneal temperature.

(v) The PPC effect decreases with increasing bias applied during the light exposure.

(vi) Large current fluctuations are observed near T_e.

Transverse Conductance

The interpretation of conductance measurements parallel to the alternating n-type and p-type layers is complicated by the fact that the applied field is never truly parallel to the layers. As explained in Section 2, perpendicular field components easily arise because unavoidable inhomogeneities change the potential profiles of adjacent layers.[11] Even though transverse fields and currents will therefore be present, we do not expect to observe capacitance charging effects in our doping-modulated multilayers because the p-layers, although more resistive than the n-layers, are much less resistive than the silicon nitride layers of the compositionally-modulated films discussed in section 2. Consequently, the RC time constant should be much less than 1 sec. In order to separate the effects caused by the two field orientations we began to study the field dependence of the transverse conductance of our doping-modulated multilayers.[7]

Fig. 10 shows that bias-cooling produces different conditions which cannot be distinguished in the ohmic region at low voltages but which yield strikingly different G(V) in the high voltage non-ohmic region.[7] A non-ohmic high conductance branch is observed when there is no bias during cooling, $V_c = 0$. The rise in excess conductance begins at larger measuring voltages V_m when the cooling bias V_c is increased. The onset occurs roughly when $V_m = V_c$.

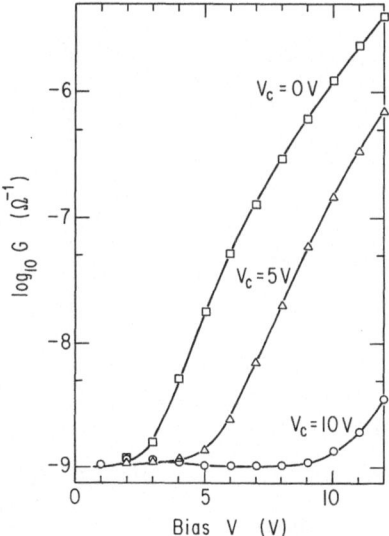

Fig.10

Transverse conductance of doping-modulated multilayer at $24^o C$ as a function of measuring voltage after bias-cooling with different V_c. After Ref. 7.

By applying a measuring voltage V_m to a sample, that has been bias-cooled with V_c, we create a metastable state when $V_m \neq V_c$ which is the same situation as encountered in the coplanar case. The $V_c=0$, $V_m=10V$ high G state is metastable and tends to relax to the $V_c=10V$, $V_m=10V$ low G state. This relaxation is exceedingly slow at room temperature and progressively faster as the temperature is increased. This results in the annealing curve of Fig. 11. The equilibration temperature is here $T_e \sim 160^\circ C$.

A large persistent photoconductance, PPC, was also observed in the transverse direction.[7] Its magnitude and bias dependence changes with the cooling-bias V_c. The effect of a bias V_e applied during light exposure on the magnitude of the PPC is considerably larger in the transverse than in the longitudinal direction. The exponential decrease is shown in Fig. 12 where J_{oe} is the PPC current with $V_e=0$. The quenching of PPC by a transverse bias indicates the importance of the built-in junction fields for creating and retaining the excess conductivity. A bias of about 0.6 V per junction practically prevents the PPC effect. The bias is close to the potential drop of 0.4 eV estimated from the difference in the conductance activation energies E_a perpendicular and parallel to the layers. The quenching effect is less pronounced when the field is applied parallel to the layers and begins only with the nonlinear I-V behavior. This might be expected because only the transverse field components associated with inhomogeneities should diminish the PPC.

A summary of these preliminary results is the following:

(i)   A transverse bias V_c applied while cooling from $T>T_e$ produces different states which become frozen-in below T_e.

(ii) Below T_e these bias-cooled states become metastable when $V_m \neq V_c$.

(iii) At low voltages the conductances of different bias-cooled states are the same.

(iv) The conductance rapidly increases when $V_m>V_c$.

(v)   The transverse PPC conductance decreases exponentially with increasing exposure bias V_e.

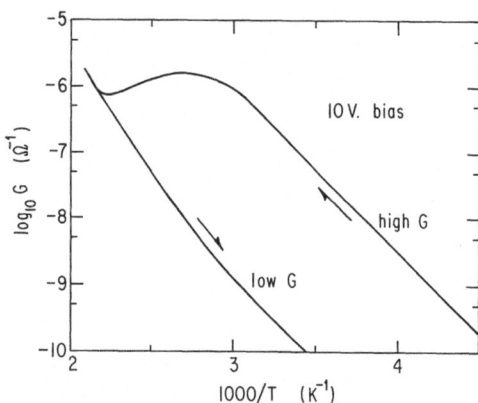

Fig.11   Annealing of the high G metastable state that was obtained by $V_c=0$ cooling and applying at $24^\circ C$ a bias $V_m=10V$. After Ref. 7.

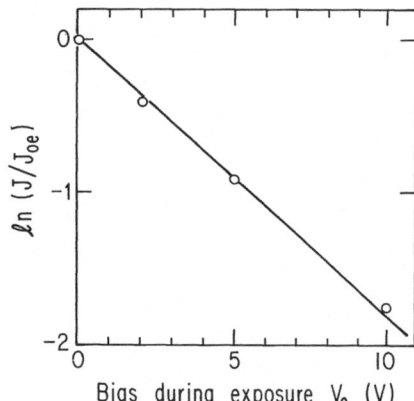

Fig.12   Decrease of PPC in the transverse direction with bias applied during light exposure. After Ref. 7.

20

CONCLUDING REMARKS

Even though we investigated semiconducting films whose heterogeneous structures have a very simple geometry, we found a variety of complex nonlinear current-voltage characteristics and interesting bias- and light-induced metastable states. The two kinds of amorphous semiconductor multilayers considered here contain separately the two dominant heterogeneities expected in more complicated materials: spatial variations in composition and internal fields.

In our compositionally-modulated multilayers, in which the semiconducting layers are interleaved with insulating layers, we believe the metastable conductance states arise from space charges that are induced by transverse field components. The experimental data agree qualitatively with this model. An analytical solution of the problem seems remote however since we would need to know the conductance inhomogeneities of the semiconducting layers. Having that, the problem is then similar to a two-dimensional field-effect transistor in which the spatial variations in the source-drain potential profile act as the gate potential.

In the doping-modulated multilayers we find new metastable states which depend on the bias applied during cooling. These metastable bias-cooled states exhibit very different nonlinear current-voltage characteristics except at small biases where their conductances are practically the same. This differs from the light-induced persistent photoconductivity where a metastable change of both the low bias and the high bias conductance is observed. The RC time constants in these multilayers are too short to explain the very slow relaxations of the metastable states. There might be a relation between the bias-induced and light-induced metastable conductance state and the glass-like behavior of the bonded hydrogen network[19] in a–Si:H that mediates the equilibration of defects and dopants in these materials. We are studying this at present.

The field quenching of PPC suggests that this effect requires fluctuating internal electric fields. These occur whenever defects of any kind or interfaces become charged. Indeed persistent photoconductivity effects have been observed not only in npnp doping-modulated multilayers structures but also in compositionally modulated multilayers,[20] in compensated a–Si:H films,[21] and in p-type films having a negatively charged oxide layer.[22]

Amorphous semiconductor multilayers exhibit electronic properties not observed in homogeneous materials. Their simple structure and the large choice of materials that can be incorporated thus allow control over a wide range of phenomena.

ACKNOWLEDGEMENTS

H. Fritzsche wishes to thank H.K. Henisch for many years of friendship. We admire the immense breadth of his intellectual and artistic life and the highly professional level of all of his pursuits. This work was supported by NSF DMR-8504717 and the Materials Research Laboratory which is funded by the National Science Foundation. Part of the research received scholarship funds from Kanegafuchi Chemical Industries, Co. and from the Korea Science and Engineering Foundation.

REFERENCES

1. H. K. Henisch "Rectifying Semiconductor Contacts" (Clarendon Press, Oxford 1957).
2. H. K. Henisch, "Semiconductor Contacts, An Approach to Ideas and Models", (Oxford University Press, 1984).
3. M. K. Sheinkman and A. Ya. Shik, Sov. Phys. Semicond. 10 (1976) 128.
4. J. Kakalios and H. Fritzsche, Phys. Rev. Lett. 53 (1984) 1602.
5. J. Kakalios, H. Fritzsche and K. L. Narasimhan, AIP Conf. Proc. 120 (1984) 425.

6. S.-H. Yang and H. Fritzsche, J. Appl. Physics, to be published.

7. J. Takada and H. Fritzsche, Proc. of the 12th ICALS, Prague, Czechoslovakia, Aug. 24-28, 1987, edited by M. Matyás, J. Kočka and B. Veličky (Special issue of J. Non-Cryst. Solids).

8. B. Abeles and T. Tiedje, Phys. Rev. Lett. 51 (1984) 2003.

9. J. Kakalios, H. Fritzsche, N. Ibaraki and S. R. Ovshinsky, J. Non-Cryst. Solids 66 (1984) 339.

10. S. C. Agarwal and S. Guha, Phys. Rev. B 31 (1985) 5547.

11 H. Ugur, Phys. Rev. B 34 (1986) 2576.

12. S. C. Agarwal and S. Guha, J. Non-Cryst. Solids 77/78 (1985) 1097.

13. J. Kakalios, Phil. Mag. B 54 (1986) 199.

14. H. Fritzsche, Mat. Res. Soc. Symp. Proc. 77 (1987) 29.

15. L. Ley in "Amorphous Semiconductors", eds. H. Fritzsche, D.-X. Han and C. C. Tsai (World Scientific, Singapore, 1987) p. 257.

16. Ch. Lee and S.-H. Choi in Proc. KOSEF/NSF Joint Seminar, The Physics of Semiconductor Materials and Applications, Eds. Ch. Lee and W. Paul (Korea Advanced Inst. of Science and Technology, Seoul, 1987) p. 17.

17. J. Kakalios, R. A. Street and W. B. Jackson, Phys. Rev. Lett. 59 (1987) 1037.

18. R. A. Street, J. Kakalios, C. C. Tsai and T. M. Hayes, Phys. Rev. B 35 (1987) 1316.

19 J. Kakalios and R. A. Street, Proc. of the 12th ICALS, Prague, Czechoslovakia, Aug. 24-28, 1987, edited by M. Matyás, J. Kočka and B. Veličky (Special issue of J. Non-Cryst. Solids).

20. K. J. Chen and H. Fritzsche, "Amorphous Semiconductors", Eds. H. Fritzsche, D.-X. Han and C. C. Tsai (World Scientific, Singapore, 1987) p. 243.

21. H. Mell and W. Beyer, J. Non-Cryst. Solids 59/60 (1983) 405.

22. B. Aker and H. Fritzsche, J. Appl. Phys. 54 (1983) 6628.

RECENT DEVELOPMENTS IN OVONIC

THRESHOLD SWITCHING DEVICE TECHNOLOGY

Roger W. Pryor

Energy Conversion Devices, Inc.
1675 West Maple Road
Troy, Michigan 48084

ABSTRACT

The first commercial Ovonic threshold switches were discrete devices with carbon electrodes in a DO-7 package. The threshold switching materials used in these devices were originally made for AC application and then adapted for DC operation by material changes such as the addition of selenium to the tellurium-based, stable chalcogenide materials. Threshold switches were successfully made in thin film form for experimental purposes.

In order to make a thin film threshold switch suitable for production processing, since the chalcogenide materials are moisture and oxygen sensitive as in the original devices, passivation and material stabilization techniques needed to be developed.[*] The development of this thin film device technology required a systems approach which continued the use of carbon electrodes at the glass surface and, at the same time, was amenable to use of photolithographic patterning and dry etching. It required the development of a conformal, inorganic passivation layer. That was achieved, assuring inherent high stability and long life. The switching material in the device has

---

[*] Moisture and impurities are the ordinary problems of any semiconductor device--crystalline or amorphous.

also been modified to incorporate hydrogen, which significantly improves the stability and uniformity of the resulting switches.

As a prototype production process, the new, all-thin-film Ovonic Threshold Switch (OTS) using the hydrogenated chalcogenide material shows device yields as high as 90+ percent. The threshold voltage variability, measured on devices made using the same conditions for individual devices under worst-case AC conditions, is typically less than 10 percent. Threshold voltage variability for the same devices, measured from device to device using the same conditions, is typically less than 5 percent.

The development of this new device technology has resulted in the ability to fabricate high yield, DC-stable, thin film threshold switches with a first-fire voltage that is substantially equal to the final threshold voltage.

INTRODUCTION

Before 1968, there was little activity in the electronically active amorphous materials area. At that time, a landmark paper on threshold switching was published by Ovshinsky [1] which resulted in a rapid expansion in the study and use of amorphous materials. [2-7] Subsequent studies of factors relating to the mechanism of threshold switching have created an extensive and varied literature. A large number of mechanisms have been postulated ranging from purely thermal to purely electronic. The experimental evidence has supported the original conclusion that the mechanism of switching is electronic in nature. A detailed analysis of the particular mechanisms is reviewed elsewhere. [8]

This paper presents the latest results of an ongoing device processing technology and switching materials development program. [9] That program has facilitated the development of a method of fabrication of a new class of thin film, DC-stable, high current Ovonic Threshold Switches using a hydrogenated chalcogenide switching material. Additionally, these devices can be fabricated so that the threshold voltage is the same for the first pulse (first-fire) and for all other pulses. In order to place these changes in perspective, the characteristics and attributes of threshold switches made by previous technologies will be reviewed briefly.

HISTORICAL BACKGROUND

Initially, typical device structures were simple and limited to those that could be made with either mechanical masking steps or no masking steps at all. Examples of such devices are shown in Figures 1a and 1b. The device shown in Figure 1a is comprised of two electrically conducting rods that have received a deposit of the selected switching material over half the circumference of the rod and over a finite portion of the rod's length. The area of deposition on the rod was defined by mechanical means. The device is formed by placing two such rods against each other, under slight pressure, with the switching material on each rod in contact with the switching material of the other rod. Figure 1b is similar to 1a except that instead of using the

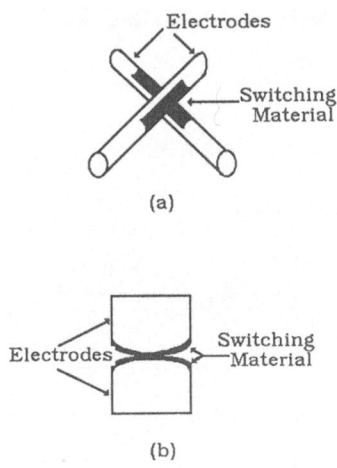

Fig. 1 Vertical Threshold Switch Geometrics

side of the rods, the deposition surfaces are spherically polished rod ends, which then allows the direction of the mechanical pressure to be applied perpendicular to the point of contact.

The original OTS work established clearly that the only highly stable electrodes were carbon and that other types of metallic electrodes were chemically interactive with the switching material. For Ovonic memory materials, platinum silicides proved suitable. It was also determined that oxygen and moisture would deteriorate the materials. The best early devices were therefore encapsulated to prevent contamination.

Figure 2a shows an example of an early lateral threshold switching device. The large electrode separations required by mechanical masking resulted in high threshold voltages. The high threshold voltages resulted in a large stored charge in the off-state. The net result was a large energy deposition in the thin films each time the device fired. Recent findings indicate that there may be a minimum thickness rule required in the fabrication of lateral devices that does not apply to vertical devices. This minimum thickness condition arises from the boundary

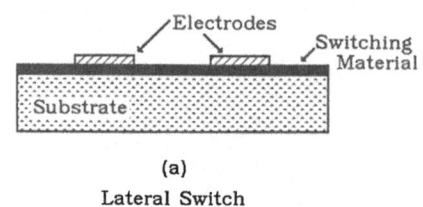

(a)
Lateral Switch

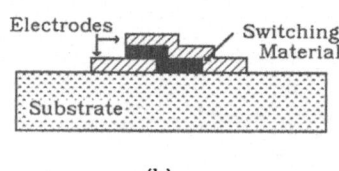

(b)

Vertical Switch

Fig. 2   Thin Film Threshold Switches

conditions associated with filament formation. That
hypothesis is presently in the process of exploration.

Figure 2b shows a typical, early, mechanically masked,
thin film, vertical switching device. These devices
suffered all the same instabilities as the previously
mentioned devices except the high voltage related
problems. They did, however, suffer instability problems
due to lack of encapsulation and also due to edge
definition problems.

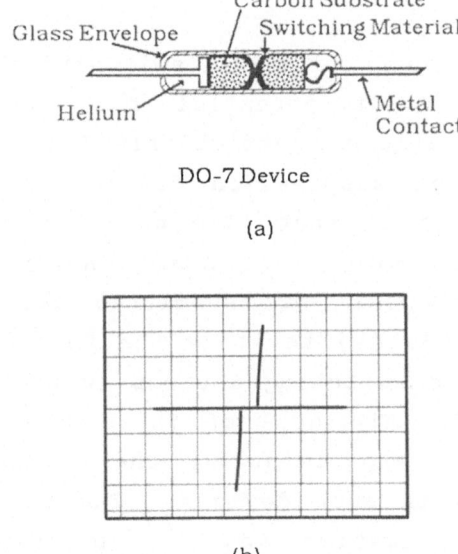

DO-7 Device

(a)

(b)

Fig. 3   DO-7 Threshold Switch and I/V Curve

The commercial Ovonic Threshold Switch device
fabrication process was accomplished by the deposition of a
chalcogenide glass (Te(39) As(36) Si(17) Ge(7) P(1)) onto a
pair of spherically formed pyrolytic graphite electrodes as
shown in Figure 1b. These electrodes were then

hermetically sealed in a glass envelope with a dry helium atmosphere inside the envelope as shown in Figure 3a. This type of technology was widely used at that time to package crystalline silicon diodes and was called a DO-7 package. Threshold switches packaged in that way, made with the materials and the technology available in the early 1960's and 1970's, were AC-stable, and with the proper material modifications, DC stable in the pulse mode. They could handle currents in the tens of milliampere range and under pulsed conditions could handle much larger currents. They have been demonstrated to have a long shelf life (15 to 20 years as measured). Figure 3b shows a typical I/V curve, as measured, using one such DO-7 packaged device. Based on the successes of the DO-7 devices, the new all-thin-film configurations discussed below were designed.

The DO-7 packaged threshold switching device has, in the past, shown great potential for use in transient suppression applications, particularly EMP. These devices and other transient suppression devices were reviewed in July 1980, [10] but at that time no further developmental action was forthcoming. A Small Business Innovation Research (SBIR) proposal was submitted early in 1985 and awarded late in 1985 [9] to study the feasibility of using thin film threshold switching devices to protect integrated circuits from EMP. The goal of Phase I of that contract was to demonstrate experimentally the current levels that would be required in such devices. The uniqueness of the OTS for that application was its picosecond switching speed, its high current density in the on-state (10^4 amps/cm^2) and the ability of the filament to expand and contract to keep a constant current density. It has the highest observed current density of any solid state semiconductor switching device and the ability to carry exceptionally large transient currents by virtue of its expanding filament.

The successful efforts of Phases I and II [9] form the basis of the presentation during the remainder of this paper. The following results are comprised mainly of experimental accomplishments and observations.

28

The research results presented in this report are for work done with vertical conduction devices of the types shown in Figures 4 and 5. Both mesa (Figure 4) and channel type devices (Figure 5) are fabricated in a serpentine geometry (Figure 6) to maximize thermal transport from the filamentary conduction region. The mesa type of device is fabricated by depositing all the active device layers (electrode, switching glass, electrode) sequentially

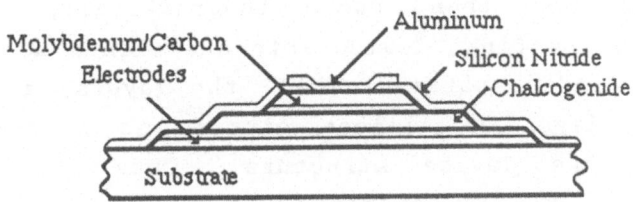

Fig. 4   Mesa Geometry for a Thin Film Threshold Switch

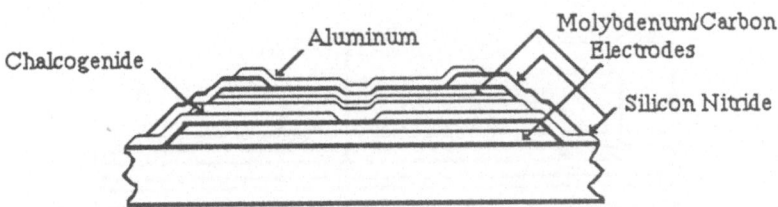

Fig. 5   Channel Geometry for a Thin Film Threshold Switch

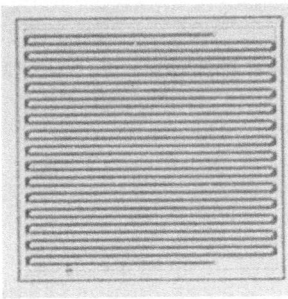

Fig. 6   Large Area Channel Geometry Ovonic Threshold Switch

without breaking vacuum. It should be noted that these devices use a two-layer electrode structure with carbon adjacent to the switching glass and molybdenum on the outside. This structure is observed to give the best electrical contact with a minimum of physical interaction between the electrode and the switching glass.

OTS devices can be made on either a conductive or an insulating substrate as long as that substrate is process compatible. The choice of substrates is dictated primarily by the device's current capability and the amount of thermal sinking needed for proper functioning of the device. Once all the layers are deposited, each layer (Figure 7) is then photolithographically patterned, transforming the final device into one with a mesa geometry (Figure 8). After patterning all the layers, a conformal, passivating layer of silicon nitride is applied to the outside of the device structure. Once the device is passivated, a via is opened to the upper electrode and contact metallization applied to form the final device (Figure 4).

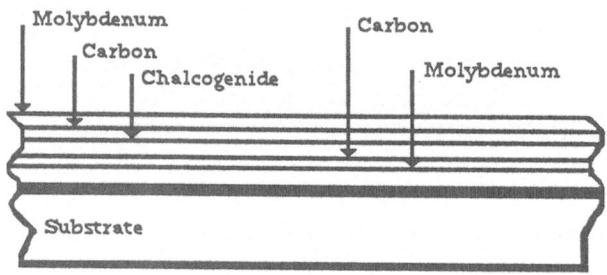

Fig. 7   Five Layer Deposition for Mesa Geometry

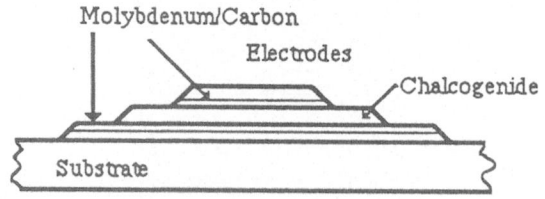

Fig. 8   Etched Five Layer Deposition for Mesa Geometry

30

The device shown in Figure 5 is termed a channel device. The channel device differs from the mesa device primarily in the method by which the active current conduction channel is defined. This, of course, changes the processing sequence. The processing sequence for this device requires deposition of the lower electrode and then deposition of an insulating layer (Figure 9). The insulating layer is then patterned to form conduction channels (Figure 10). After patterning, the switching glass and upper electrode layers are deposited without breaking vacuum (Figure 11). The resulting structure is then patterned to form discrete devices (Figure 12). The devices are passivated, vias are opened and then contact metallization is applied (Figure 5).

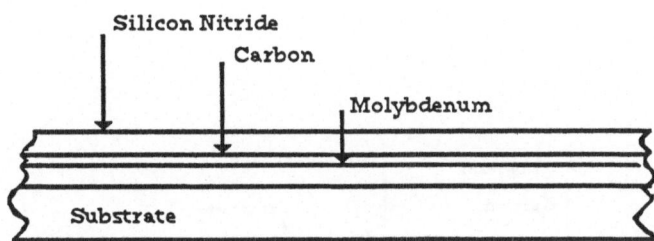

Fig. 9   Three Layer Deposition for Channel Geometry

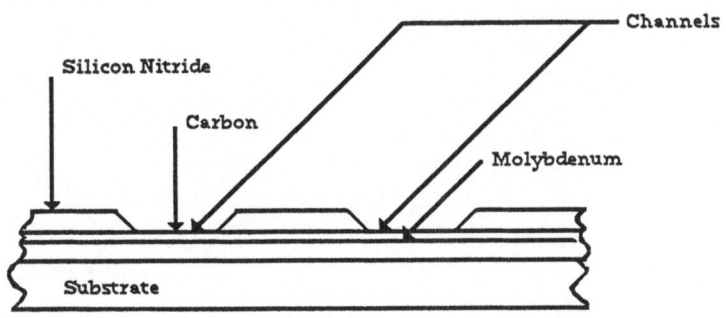

Fig. 10   Three Layer Deposition with Channels

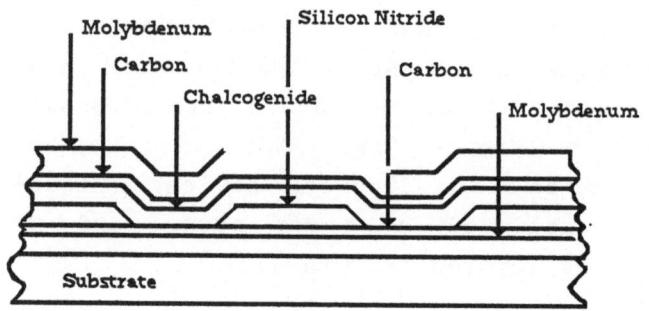

Fig. 11   Six Layer Deposition with Channels

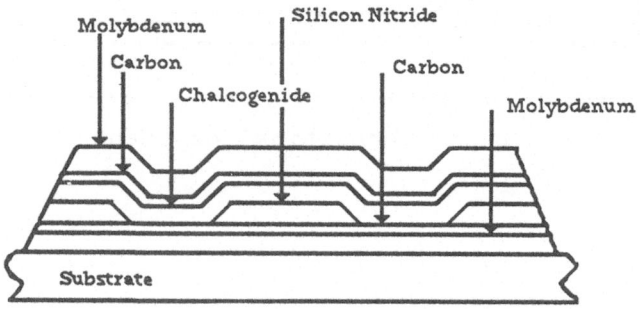

Fig. 12   Etched Six Layer Deposition with Channels

The mesa type device structure has the advantage that it can be fabricated without breaking vacuum during active layer deposition. This limits the potential contamination introduced into the bulk of the device to only that which is present in the vacuum system. This does not eliminate edge and surface contamination. The edges and surfaces of the device may also be subject to contamination during the subsequent processing. If those surfaces are not properly treated prior to application of the passivation layer, edge and surface contamination can create both electrical and mechanical problems. The successful process will be summarized below. The detailed solution to each of the potential problems will not be treated here, but will be the subject of later publications.

The channel type device structure has two potential advantages: first, in this structure heat is conducted away from the conducting channel more rapidly than in the case of the filament surrounded only by glass; second, the conducting channel geometry is smaller, allowing denser device structures. The disadvantage of the channel type device structure is that it has a high potential for contamination in the active area. This problem can be successfully solved with careful procedures for channel preparation prior to the switching glass deposition step. The approach that was used to achieve a solution to this problem is summarized below. The detailed process will be presented later in a subsequent paper.

PROCESSING TECHNIQUES

The processing techniques used to fabricate the new, thin film, high current, DC-stable, threshold switching devices are similar to those used in the crystalline silicon integrated circuit industry. They are, however, significantly modified to accommodate the unique materials utilized in these devices.

The following table shows the materials used in fabricating both a typical mesa type device and a typical channel type device. The approximate thickness of each layer of each material for that device type is shown.

Material	Thickness (nm)	
Device Type	Mesa	Channel
Molybdenum	500	500
Carbon	100	100
Silicon Nitride	0	190
Chalcogenide + Hydrogen	350	350
Carbon	100	100
Molybdenum	300	500

In both the mesa and channel device structures, the molybdenum and carbon electrode layers are deposited using a DC magnetron sputter deposition technique. In using this deposition technique, the parameters are set so that the electrode layers are deposited pinhole and stress free, a system-dependent consideration. In the case of the mesa type devices, the molybdenum and carbon layer depositions are followed immediately by the deposition of a layer of chalcogenide switching glass (Te(36) Ge(23) S(21) As(18) Se(2)), reactively RF sputtered with hydrogen, in an argon environment. The presence of hydrogen during the deposition of the chalcogenide is observed to significantly improve the quality and reproducibility of the devices. Hydrogen, in the correct amounts, causes the threshold voltage to be substantially the same for all switching events including the first switching event. Figure 13 shows a comparison of a hydrogenated and a non-hydrogenated OTS. As can be seen, the hydrogenated device (Figure 13b) shows significantly better performance than the unhydrogenated device (Figure 13a). After the chalcogenide is deposited, the next step in fabricating a mesa device is the deposition of the carbon and molybdenum layers of the upper electrode structure on top of the chalcogenide layer without breaking vacuum. The carbon and the molybdenum layers, comprising that electrode, are DC magnetron sputtered under low stress conditions.

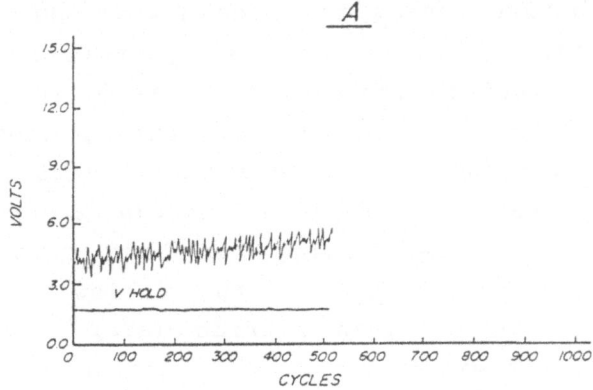

A

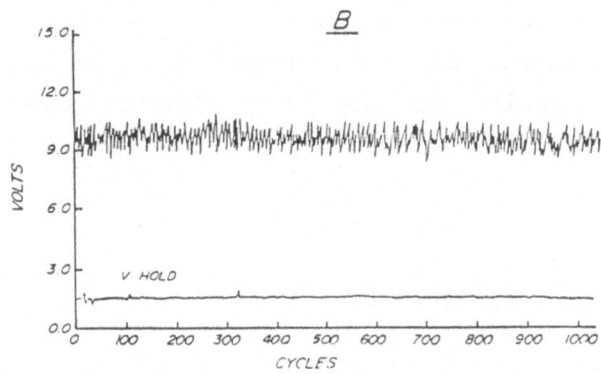

B

Fig. 13　Thin Film OTS Device Stability
A.　Without Hydrogen
B.　With Hydrogen

In the case of the channel type device, a layer of CVD silicon nitride is deposited on top of the free carbon surface of the lower electrode. The CVD silicon nitride is then photolithographically patterned and dry etched to open conductive channels to the carbon layer. As a result of this processing, residues can be left in the channel area, which will diffuse into and destabilize the switching glass under high field/current conditions. This problem is solved by cleaning the channel with a sputter etch immediately before subsequent depositions. Once the channel has been cleaned, the chalcogenide and upper electrode layers are deposited sequentially without breaking vacuum. As with the mesa type device, the chalcogenide layer is reactively RF sputtered with hydrogen. The carbon and molybdenum layers are DC magnetron sputtered.

35

Once the active layers for each device type have been deposited, then each layer is photolithographically patterned and etched to achieve the final device structure, as shown in Figures 4 through 12. During those etching steps, the molybdenum and chalcogenide layers are liquid etched. The carbon and silicon nitride layers are dry etched. After all the etching has been completed, the final device is passivated with a low temperature, reactively sputtered silicon nitride layer. The nitride layer is then patterned for a via etch and dry etched. After etching, a layer of application dependent contact metallization is applied to the upper surface, patterned and etched. If the substrate is conductive (e.g., degenerate silicon), an application dependent metallization may also be applied to the back side of the substrate. Typical metals are aluminum for wirebonding and copper for soldering.

ELECTRICAL CHARACTERISTICS

The electrical characteristics of the new, thin film OTS are similar to those of the previously made threshold switching devices. They are, because of their packaging and configurations, fundamentally improved in a number of aspects. The new devices are both AC- and DC-stable and, due to new electrode configurations, carry significantly higher overall currents than previously available thin film devices. Figure 14b shows, through the use of a composite picture, the DC-stable current/voltage characteristic curve of a thin film OTS. This picture was taken by a multiple exposure technique in which each spot represents one DC-operating point. The same device is shown, for comparison, under AC-operating conditions in Figure 14a. In these figures, the device is shown operating at a peak current of 80 milliamperes in both the AC- and DC-operating conditions.

A

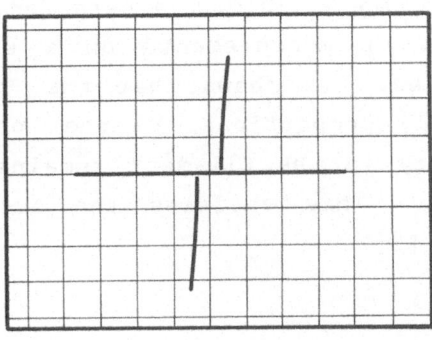

B

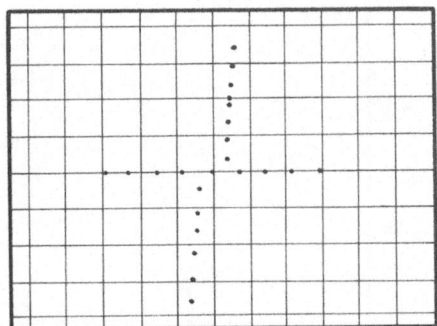

Fig. 14   Thin Film, High Current OTS Device
          A.   AC I/V Curve
          B.   DC I/V Curve (composite picture)

One example of the higher current capability of these devices is shown in the next graph. (Other larger devices have been measured at currents as high as 1.8 amperes.) Figure 15 shows current/voltage characteristic of a 1250 sq. micron channel type device. This device achieves "pore" saturation at approximately 250 milliamperes as shown by the kink in the curve. The current density calculated at that "pore" saturation point (the kink in the curve) is approximately 20,000 amperes/sq.cm. Figure 16 shows the results of measurements on a set of different "pore" size devices. It shows that the "pore" saturation current is directly proportional to the "pore" area. Thus the current density in the filament remains constant below "pore" saturation. This verifies the original expanding filament finding. [1]

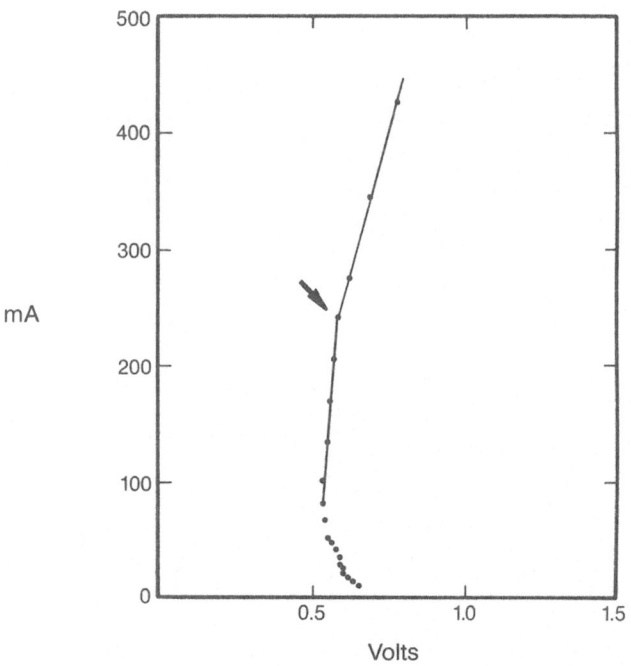

Figure 15   Current vs. voltage characteristic curve for stripe geometry device. Device Area = 1.25×10^{-5}. The pore saturation point indicated by the arrow is at I = 240mA. The initial series resistance is 0.27Ω and the resistance after pore saturation is 1.1Ω.

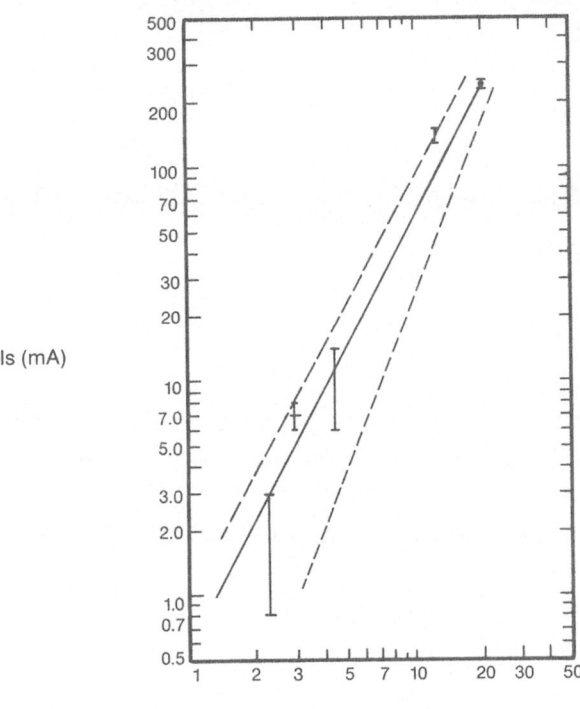

Is (mA)

Effective Radius (µm)

Figure 16   Saturation   Current   vs.   Effective   Pore
Saturation Radius.   I Pryor and Henisch [J.
Non-Cryst. Solids 7 (1972) 181].   Dashed lines
indicate data from Petersen and Adler [J. Appl.
Phys. 47 (1976) 256].   Solid line has a slope
of 2 and passes through the current data (solid
bars).   Current density = 1.9×10^4 A/cm^2.

The devices have been measured for threshold stability
under both AC (alternating half-sinusoids) and DC (fast
rectangular pulse) conditions.   Figure 17 shows a typical
AC, threshold voltage versus number of cycles graph.   The
same device was also measured under DC conditions (Figure
18).   It can be easily seen that the variability of the
threshold voltage, even under worst-case AC conditions, is
small.   Under DC conditions, the threshold voltage
variability is almost zero.

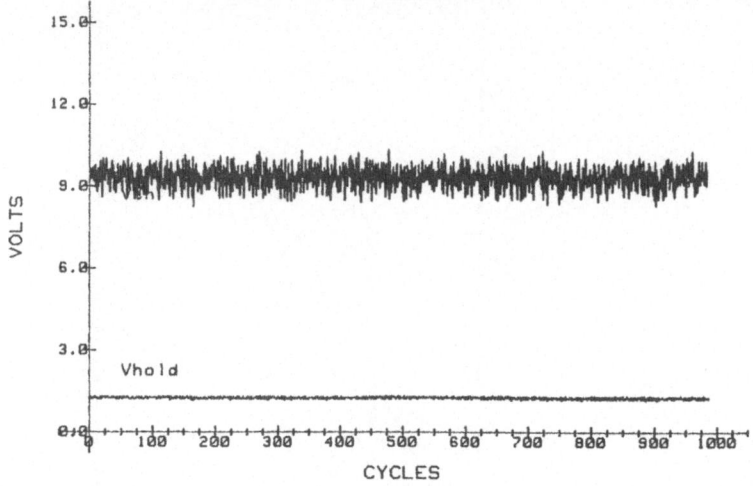

Fig. 17   AC Test of OTS Threshold Voltage Stability

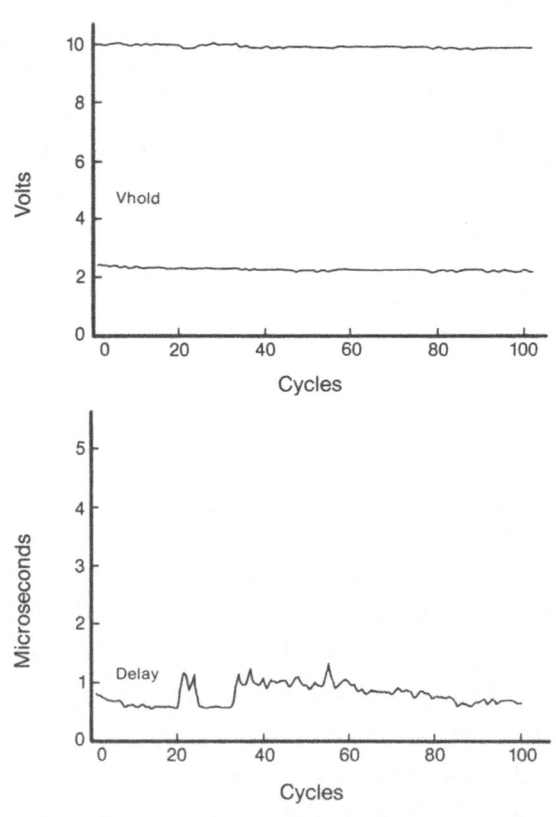

Fig. 18   DC (Pulse) Test of OTS Threshold Voltage Stability

Figure 19 shows a graphical comparison of the the first-fire voltage, threshold voltage and threshold voltage stability for a set of devices on the same substrate. It can be easily seen that the first-fire voltages are approximately equal to the subsequent threshold voltages, that the threshold voltage deviates little from device to device and that the threshold voltage for each device deviates little from event to event. Yield on properly processed devices typically has a value of 90+ percent.

THEORY

Hydrogen is well-known to be an electronically active material. In amorphous tetrahedral films, it affects the density of states [11] and has been shown to significantly modify the properties of trapping sites in thin, insulating films. [12-14] In non-switching arsenic-selenide materials, Fritzsche found hydrogen had no impact on the electrical properties. [15,16] The important factor here is that this is the first known occurrence of a chalcogenide switching material that not only has hydrogen

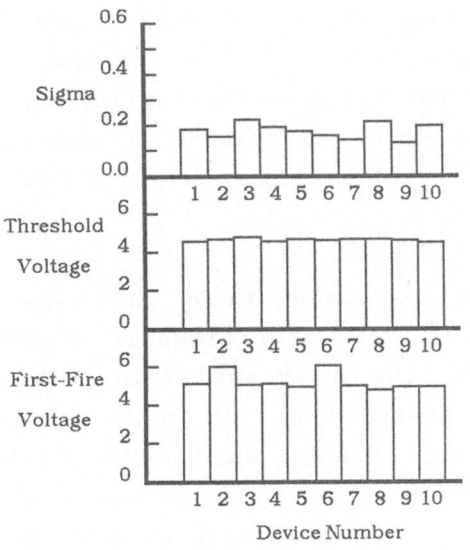

Fig. 19   OTS Device Reproducibility

41

present in the film but where that hydrogen is also electronically active. A detailed examination of the role of hydrogen in these materials and its effect on switching will be presented in a later publication.

Numerous measurements have been made on various hydrogenated and unhydrogenated devices as shown in the earlier text. These measurements show a consistent reduction in the width of the threshold voltage distribution, measured under AC conditions, for hydrogenated devices. An increase in the bandgap in chalcogenide specimens, which are normally opaque when unhydrogenated, is shown by an increased optical transparency of the specimens in the visible region when they are made to include a few atomic percent hydrogen.

SUMMARY and CONCLUSIONS

A new technology has been developed for the fabrication of Ovonic Threshold Switches. This technology includes the development of both new device structures and new hydrogenated, chalcogenide OTS device materials. This technology yields devices that are thin film, high current devices that are both AC- and DC-stable. The devices show a first-fire voltage that is substantially equal to the subsequent threshold voltage. This technology also allows process device yields in the 90+ percent range of devices that have threshold voltage variability on the order of 10 percent, measured under worst-case AC conditions. Device to device threshold voltage variability has been measured at the 5 percent level.

The OTS devices made with this new technology and advanced materials are significantly improved over those devices made with previously available technology. These new devices, due to their thin film configuration and excellent stability and yield, present an opportunity for the OTS to be generally utilized in electronics design. Obvious applications for these devices are ESD/EMP transient suppression for integrated circuits. Other applications for the device are in oscillators, power supplies, etc.

Acknowledgments

I would like to thank S. R. Ovshinsky for his support, advice and for his historical insight during the development of the new thin film OTS. I would also like to thank him for his review of this manuscript.

I am very grateful to Patrick J. Klersy, Napo Formigoni and Jerry Piontkowski for their many contributions to the success of the OTS project. I am also thankful for the helpful advice of H. K. Henisch and for the hydrogen measurements of W. A. Lanford.

References

1. N. F. Mott and E. A. Davis, "Electronic Processes in Non-crystalline Materials," Second Edition, Oxford University Press (1979).
2. J. M. Ziman, "Models of Disorder," First Edition, Cambridge University Press (1979).
3. D. Adler and B. B. Schwartz, Institute for Amorphous Studies Series, Plenum Press (1985 -).
4. e.g., J. Appl. Phys.
5. e.g., IEEE Transactions on Magnetics.
6. e.g., J. Non-Crystalline Solids.
7. S. R. Ovshinsky, Phys. Rev. Lett., 21;20:1450 (1968).
8. D. Adler, H. K. Henisch, N. F. Mott, Rev. Mod. Phys., 50;2:209 (1978)
9. U.S. Army Medical Corp., SBIR Contract No.DAMD17-86-C-6158.
10. Report on the "First Workshop on Threshold Switching," U.S. Army Research Office, Workshop Sponsor: C. Beghosian, July 1980.
11. D. Adler and H. Fritzsche, "Tetrahedrally-Bonded Amorphous Semiconductors," Institute for Amorphous Studies Series, Plenum Press (1985).
12. R. W. Pryor, J. Appl. Phys., 52;5:3702 (1981).
13. R. W. Pryor, IEEE Electron Device Lett., EDL-6;1:31 (1984).
14. R. W. Pryor, IEEE Electron Device Lett., EDL-6;7:377 (1985).
15. V. Smidand and H. Fritzsche, Sol. St. Comm., 33;7:735 (1980).
16. H. Fritzsche, V. Smid, H. Ugur, P. J. Gaczi, J. dePhysique, Colloque C-4, Supplement 10, 42:699 (1981).

AMORPHOUS CHALCOGENIDE MICROWAVE SWITCHES

D. N. Bose

Materials Science Centre
Indian Institute of Technology
Kharagpur, India 721302

ABSTRACT

Amorphous chalcogenides in the form of bulk or thin films have been shown to act as microwave switches in the X-band. The salient features of these switches viz. variation of attenuation with driving current and frequency response are described. These switches are much simpler than conventional p-i-n diodes and have the advantage of displaying memory type behavior. Their characteristics also throw light on the electrical properties of chalcogenides at high frequencies.

INTRODUCTION

Switching in amorphous chalcogenides has been widely studied since the first report by Ovshinsky.(1) Classification into 'threshold' and 'memory' type switches was described by Henisch (2) on the basis of the location of the composition with respect to the glass-forming region. There ensued considerable debate regarding the nature of the switching phenomenon, whether thermal or electronic. While memory switching was proved to be essentially thermal in nature involving transitions between ordered and disordered phases, the mechanism of threshold switching was shown to depend on the thermal boundary conditions. Switching delay (3) was a phenomenon much studied in an effort to resolve this question, the delay time τ_d being found to be proportional to exp $\{-(V-V_{th})\}$ where V = applied voltage and V_{th} = threshold voltage for switching.(4)

A more fundamental aspect involves the high frequency conductivity of amorphous chalcogenides which was shown by Pollak (5) to increase as ω^n. The exponent n was expected to increase with temperature and decrease with frequency. A different model based on bipolaronic hopping was proposed by Elliott (6) in which the temperature and frequency dependences of n were opposite in nature. The present experiments were motivated to examine the frequency dependence in the microwave region and also to explore the possibility of microwave switching. Experiments were thus carried out on both threshold as well as memory type chalcogenide semiconductors as described earlier.(7,8) A brief description of these experiments is given with some new explanations for the observed behaviour.

EXPERIMENTAL
Multicomponent chalcogenides with the following compositions were prepared from the elements by conventional melting in sealed quartz ampules:

Threshold materials:
$Ge_{10}Si_8Te_{50}As_{32}$, $Ge_{17}Te_{55}As_{28}$,
$Ge_{15}Si_{20}Te_{40}As_{25}$, $Ge_7Si_{18}Te_{40}As_{35}$

Memory materials:
$Ge_{25}A_{25}Te_{50}$, $Ge_{15}Te_{81}As_4$, $Ge_{15}Te_{81}Sb_4$.

The conductance and capacitance of samples were measured between 100Hz – 100KHz using a GR1650 bridge (7) and also in the X-band using a microwave cavity. It was thus found that compositions such as $Ge_{17}Te_{55}As_{28}$, $Ge_{15}Si_{20}Te_{40}As_{26}$ and $Ge_7Si_{18}Te_{40}As_{35}$ showed conductivity saturation at frequencies as low as 100KHz. The value of n decreased typically from 0.685 at 1 KHz to 0.54 at 100KHz. Over the frequency range 1 KHz – 10GHz the average value of n was 0.517. This decrease in n with increasing frequency is in agreement with the Pollak theory.(5)

Samples were used either in bulk form (0.02 cm thick) or as evaporated thin films (0.12 μm thick) as shown in Fig. 1 for experiments in microwave switching. The arrangement, depicted in Fig. 2, consisted of a Gunn oscillator source, a pulse generator with the switch mounted on a reduced height wave guide followed by a detector with output displayed on an oscilloscope. The threshold voltage for the device was typically on the range $5-10^V$ with the holding voltage being ~ 0.5V. A few milliwatts of microwave power at 9.35 GHz was modulated by the threshold device which was supplied with pulses of maximum amplitude 30^V at a frequency which could be varied between 10Hz – 100 MHz.

RESULTS AND DISCUSSION

With increasing voltage applied to the switch, the microwave signal was modulated, the detected output being 180° out of phase with the modulator signal. This is because of the reflection of the incident microwave signal by the switch when it is in a conducting state at V > V_{th}. The variation of isolation with d.c. device current is shown in Fig. 3 in which switching was found to occur at 1 mA

The loss increased from 5 dB immediatly on switching at 1 mA to 10 dB at I = 10 mA and 18 dB at I = 100 mA.

The frequency of the modulating signal was increased up to 100 MHz to find a switching delay. No delay was found even though the applied voltage was kept just above V_{th}. However the detected output decreased by 2.5 dB as the modulating frequency increased from 1 MHz and fell by 6 dB at 100 MHz as shown in Fig. 4. The switching delay if any was obviously less than 1–2 n sec in agreement with the observation of Buckley & Holmberg (9) that in devices less than 1.4 μm thick the delay was less than 0.1 n sec.

Peterson and Adler (10) had explained the electronic nature of switching in devices with adequate heat sinking

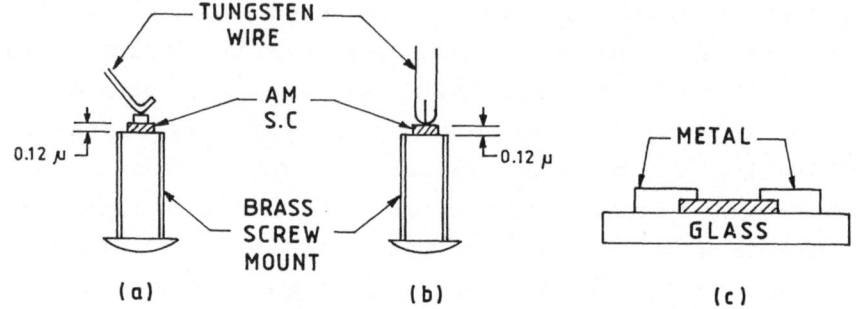

Fig. 1 Structures of amorphous semiconductor
microwave switches.

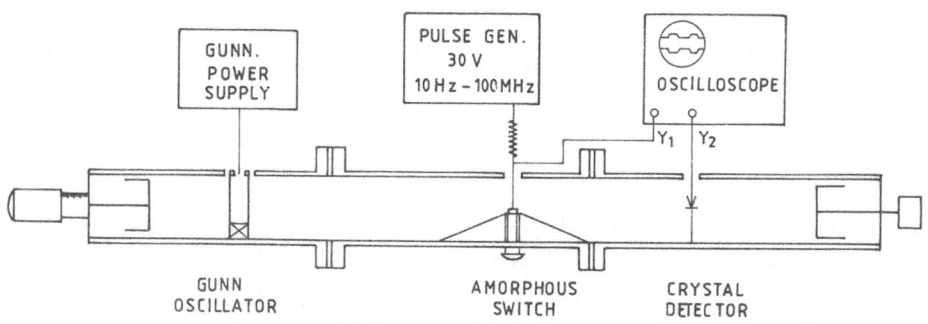

Fig. 2 Experimental arrangement for studying
microwave switching.

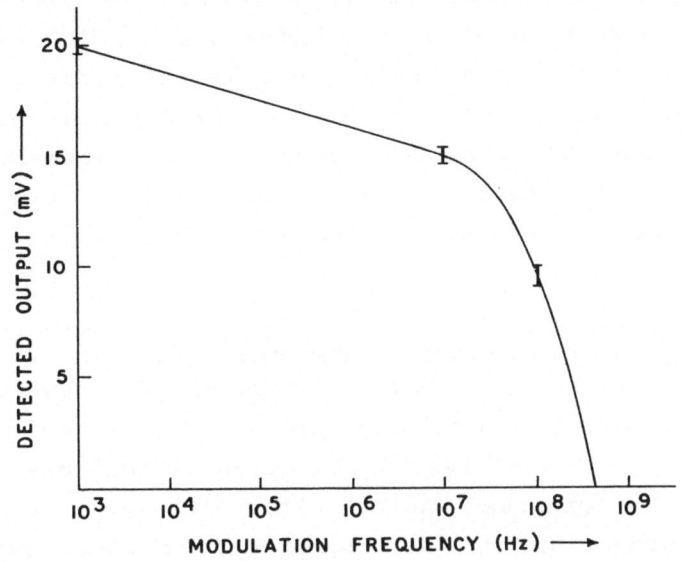

Fig. 3 Variation of loss with device current.

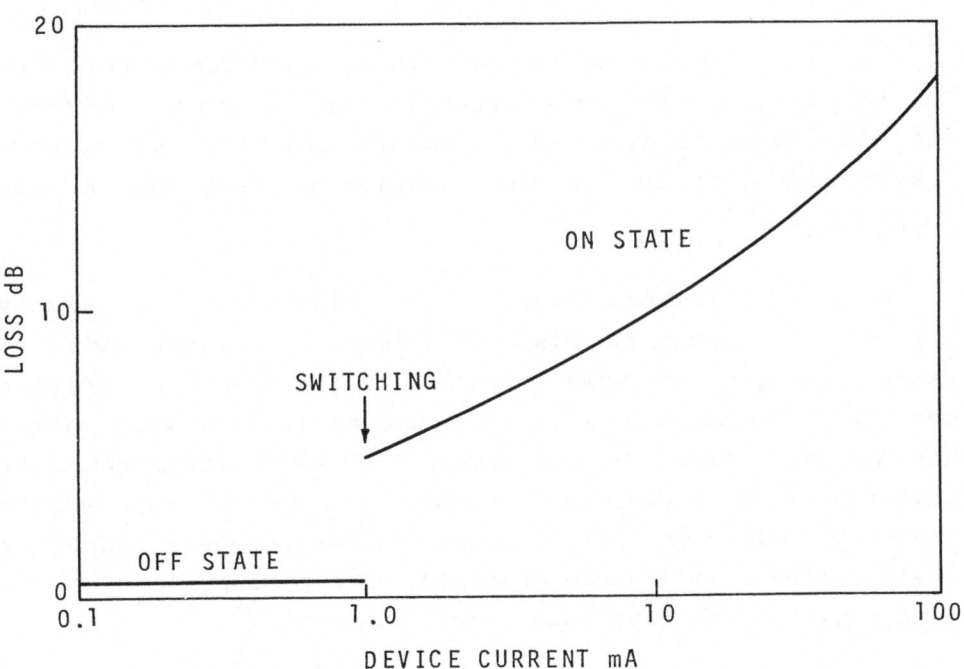

Fig. 4 Variation of modulation efficiency with
modulating frequency.

in terms of the formation of a conducting filament with constant current density. An increase in device current would then result in increase in filament radius. However such an assumption does not adequately explain the present results since the observed increase in conductance is only by a factor of 9 compared with the predicted value of 100. This follows from the device isolation relation

$$L = 20 \log_{10} \frac{G_n + 2}{2}$$

where G_n = normalised onductance and L = measured isolation loss in dB. This discrepancy had previously been explained taking into account device capacitance.(8) However, an alternative explanation which may be more realistic is proposed here. This involves space-charge limited conduction in the conducting filament due to the trap-filling as postulated earlier by Fritzsche and Ovshinsky.(11) Since $I \propto V^2$ in the case of space-charge conduction, the conductance $G \propto \frac{dI}{dV} \propto (I)^{\frac{1}{2}}$. This would explain the observed behaviour of loss with device current, since G would increase by a factor of 10 with an increase in I by a factor of 100.

The sharp decrease of modulation efficiency near 100 MHz had been attributed to transit-time effects. However the slow decrease from 10^3 – 10^7 Hz suggests that a more likely reason is due to the dynamics of trap filling and emptying.

Detailed studies were not conducted on memory switches. However it was established that such switches caused changes in VSWR with no requirement of holding current. The magnitude of the changes in loss were larger due to lower ON-state resistance. The OFF-state could be restored using a resetting pulse. In view of this unique property and the well known radiation resistance of chalcogenides, such devices could be used instead of p-i-n diodes with no holding power requirements.

50

REFERENCES

1) S. R. Ovshinsky, Phys. Rev. Lett. 21:1450 (1968).

2) H. K. Henisch, Sci, Am. 221:30 (1969).

3) H. K. Henisch and R. W. Pryor, Sol. St. Electr. 14:765 (1971).

4) F. Fritzsche, IBM J. Res. Dev. 13:515 (1969).

5) M. Pollak, Philos. Mag. 23:519 (1971).

6) S. R. Elliott, Philos. Mag. 36:305 (1977).

7) D. N. Bose and B. J. Jani, Electron Lett. 13:451 (1977).

8) D. N. Bose and B. J. Jani, Thin Sol. Films, 57:39 (1979).

9) W. D. Buckley and S. H. Holmberg, Phys. Rev. Lett. 32:1429 (1974).

10) K. E. Petersen and D. E. Adler, J. Appl. Phys. 47:256 (1976).

11) H. Fritsche and S. R. Ovshinsky, J. Non. Cryst. Solids, 2:393 (1970).

SWITCHING AND MEMORY EFFECTS IN THIN AMORPHOUS CHALCOGENIDE FILMS - THERMOPHONIC STUDIES

Melvin P. Shaw

Department of Electrical & Computer Engineering
Wayne State University
Detroit, MI 48202

We review the behavior of amorphous chalcogenide switching and memory devices, with emphasis on an experiment that measures the acoustic response of a thermophonic cell enclosing an electrically pulsed device. Here the response of the cell is measured simultaneously with the current and voltage versus time profiles. The results clearly demonstrate the fundamentally electronic nature of the switching transition, surprisingly even in thick films. Further, the results concur with a model where the sample heats during the delay time; the heating causes the fields to rearrange and a critical value is reached near an electrode, whereby a carrier generation process is encouraged.

It has been almost twenty years since the publication of S.R. Ovshinsky's paper[1] that initiated a flurry of activity in the area of amorphous materials. Since then, major progress has been made in the Physics, Chemistry and applications of these materials, particularly in the area of thin films of amorphous semiconductors[2]. Ovshinsky's paper focused on switching and memory effects in thin amorphous chalcogenides; it is the purpose of the present paper to briefly touch on the major features of the transport and device aspects of these films[3,4]. For a general overview of the process we consider Fig. 1, which is the current-voltage I(V) curve for a typical chalcogenide film (As$_x$Te$_{1-x}$, Ge$_x$Te$_{1-x}$, etc.) about 1 μm thick sandwiched between non-reactive metal electrodes.

The OFF and ON states are noted in Fig. 1, along with the dc load line, dcll. V_t, I_t is the threshold point. The Lissajous type figure (thin solid line) represents the circuit response[2], here seen as a damped oscillation as the energy is exchanged during the switching event between the local reactive components. V_h, I_h is the holding point, which is circuit dependent. There is a minimum holding current I_{hm}, below which the sample always switches back to the OFF state. There is also a current I_t, above which memory can occur. That is, if the sample is kept ON at currents above I_f for a sufficiently long time, it remains ON and follows back along the dot-dashed characteristic. A crystalline filamentary path then connects the electrodes[5]. As we know, a memory switch can readily be reset to a threshold switch by the application of a sufficiently intense and long current pulse. Figure 1 also shows a line of arrows. These represent the trajectory that is followed by the transport component of the current during the switching transition, wherein current filamentation occurs. The Lissajous type figure is shown in the plane of conduction current, i, vs. voltage, whereas the dc load line corresponds to the plane of total current I, which includes the displacement current, vs. voltage. The displacement current (capacitive discharge) plays a major role in controlling the thermal aspects of the problem, especially with regard to forming and memory effects[6,7].

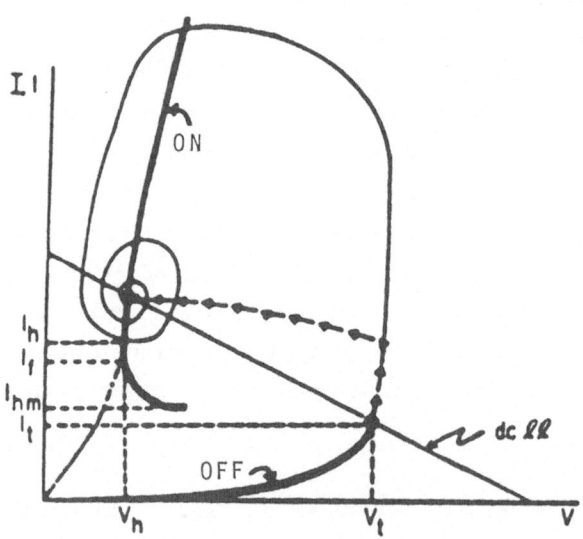

Fig. 1.   Estimated V(I) and V(i) curves for the case where the minimum current does not fall low enough to quench the filamentary inhomogeneity. (All parameters are defined in the text.) The circuit oscillations damp, a filament forms, and switching

54

Figure 2 shows the calculated[8] and experimental[9] OFF state characteristics and the switching point (where the OFF state curves terminate at high fields). In order to fit the experimental curves the heat flow equation was solved using an electrical conductivity, σ, of the following form:

$$\sigma = \sigma_o \exp\left[-\frac{\Delta E - \beta\varepsilon}{kT} - \frac{\varepsilon_o}{\varepsilon}\right],$$

where σ_o is a constant, kT is the thermal energy, ΔE is the band gap, $\beta\varepsilon$ represents a field induced decrease in the band gap, and $\varepsilon_o/\varepsilon$ represents field-induced carrier generation. It is important to note that the fit required the use of a critical electric field, ε_c, that initiated the switching transition[10]; without ε_c a good fit between theory and experiment was not obtained.

H.K. Henisch had suggested the existence of ε_c in his early work[3]. Indeed, his TONC (Transient On Characteristics) studies of the ON state with R.W. Pryor[11] were crucial to our appreciation that the maintenance of the ON state was _fundamentally_ an electronic, rather than thermal process. Figure 3 shows the essence of the TONC experiment[12] and Fig. 4 shows the results of a study by K.E. Petersen and D. Adler that elucidates the nature of the current dependence of the filamentary conducting path in the ON state[12].

The nature of the switching transition, recovery properties and specific models for the phenomenon itself has been covered in detail in prior reviews[2,3,4]. Suffice it to say that the initiation of switching in thin amorphous chalcogenide films is an electronic process. Further-

---

occurs to the ON state. The ON state is the heavy dot. As V_B is varied the dots fall on a "filament characteristic," which is sketched (dark). The switch will take place in a damped oscillatory fashion. A "spiraling in" to the ON state is fundamental to the switching process in general. In the current-time profile this appears as a damped "ringing," whose frequency is determined in part by the package capacitance and intrinsic plus package inductance.

more, the maintenance of the filamentary ON-state in threshold switches is also electronic in nature[11,13]. The switching transition occurs when a critical electric field is reached somehwere in the sample, usually near an electrode. Field-induced carrier generation then causes the charged traps to fill (neutralize). When all the traps are filled, carriers can transit the sample with an enhanced mobility and the generation rate required to keep the traps filled is reduced from its threshold value; switching then occurs.

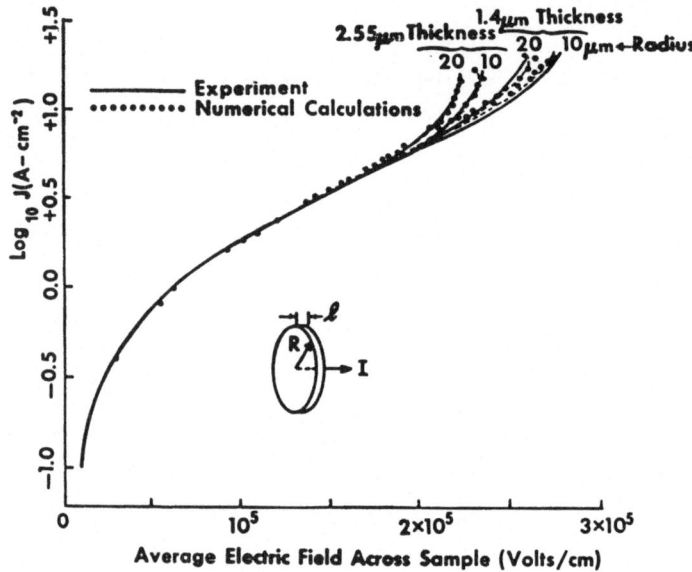

Fig. 2.   Current density vs. average electric field. Solid lines: experiment, Ref. 9. Dotted line: calculations, Ref. 8. The sample geometry is shown in the figure.

It is clear that switching in amorphous chalcogenide films, as in all other phenomena involving power dissipation, exhibits thermal effects. The basic thermal aspects of the switching process appear in the delay time, which has an appreciable thermal component, the forming process, often affected most by the capacitive discharge, and in the conventional memory effect, which is explained readily via an amorphous-to-crystalline phase transition induced at a sufficiently high temperature. The thermal effects that dominate prethreshold heating and the delay time may also help determine the primary filament nucleation site. Depending upon how the sample is driven (short or long pulses), thermal effects can either lead or follow electronic effects prior to the

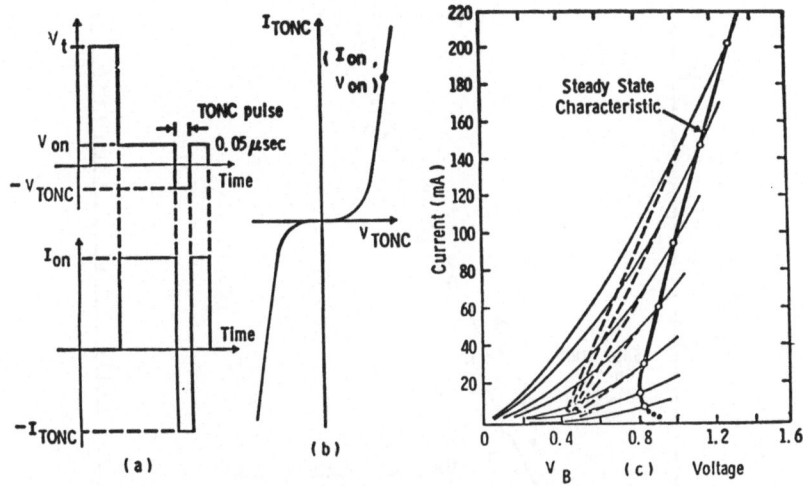

Fig. 3. The TONC experiment[11,12]. Figure (a) shows the pulse sequence used to obtain the TONC I(V) characteristic shown in Figure (b). Figure (c) shows the different TONC curves that result, depending upon the starting point (I_{on}, V_{on}) on the steady state characteristic .

switching transition, which always produces a capacitive circuit-induced thermal spike. In the remainder of this review we will discuss an experiment that very neatly separates the thermal and electronic effects: electroacoustic spectroscopy[6,7], or thermophonics.

Electroacoustic spectroscopy is akin to photoacoustic spectroscopy: a sample is optically pulse heated and the resulting thermal and acou-

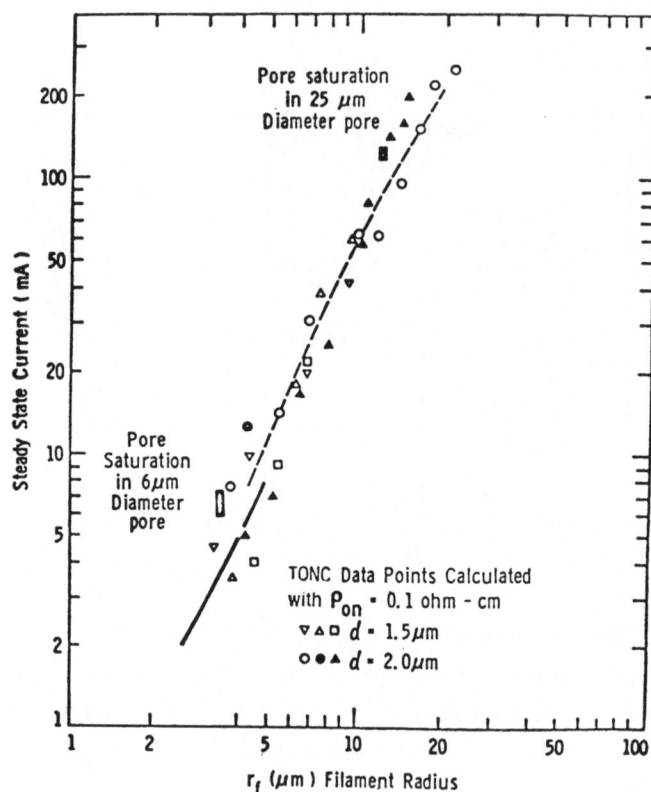

Fig. 4.   Current filament radius as determined from the TONC experiment[12] shown in Fig. 3.   The currents at which the filament fills the entire device are noted as the pore saturation values.   For currents above these values the samples exhibited non-reversible change in their properties.

stic waves emanating from the sample are detected by a miniature microphone in close proximity to the surface. In electroacoustic spectroscopy the sample is heated by an electrical pulse. In the study reported on here samples of $Ge_{15}Te8_1S_2Sb_2$ were sputter deposited through a 0.5 in. mask onto a gap-masked molybdenum coated glass substrate in a coplanar gap geometry, as shown in Fig.5. Gap widths are between 100 and 400 μm and the height of a typical sample (deposition thickness) was between 5 and 10 μm. Electrical pulses were supplied from a Cober model 606P pulse generator, primarily in a single-shot mode. The current and voltage were detected with probes and the sample encased in a thermophonic cell containing a Knowles miniature model BT-1785 condenser microphone.

Figure 5 displays a series of single-shot thermophonic signals for threshold switching events as a function of ON-state power in a sample about 100 μm widing having a threshold voltage of almost 400 V. Each trace was recorded simultaneously with the current (or voltage) time profile, where a displacement current spike always occurred at the switching transition after a delay time t_d. This capacitive discharge was mirrored in the microphone signal under certain conditions, as shown in the top trace of Fig. 5. With increasing ON-state power (1) an increase in the slope of the ON-state signal, (2) a decrease and eventual disappearance of the mirrored discharge signal, and (3) a decrease in t_d was observed. The top trace in Fig. 5 shows a typical single-shot signal under low ON-state power conditions for a first-fire event in a previously virgin (unswitched) sample. Under these bias conditions we see no evidence of forming (changes in electrical parameters due to changes in the material properties of the sample) in that the threshold voltage and OFF-state resistance remain unchanged and no morphological changes are in evidence on the surface of the sample under 40X magnification. Repetitive firings under these conditions did not produce forming, memory events or any indication whatsoever of alterations in the structural or physical properties of the sample; although, as we shall see, the capacitive discharge provides sufficient energy to raise the temperature of the sample by 200–300°C. However, as the ON-state power was increased by reducing the load resistance, the surface of the sample showed indications of morphological changes and the OFF-state resistance was reduced; the forming process was clearly in progress, and in our case it appears that the major reason for it was excessive heating. But, as stated above, the entire series of events depicted in Fig. 5 were threshold switches; memory was not observed. It

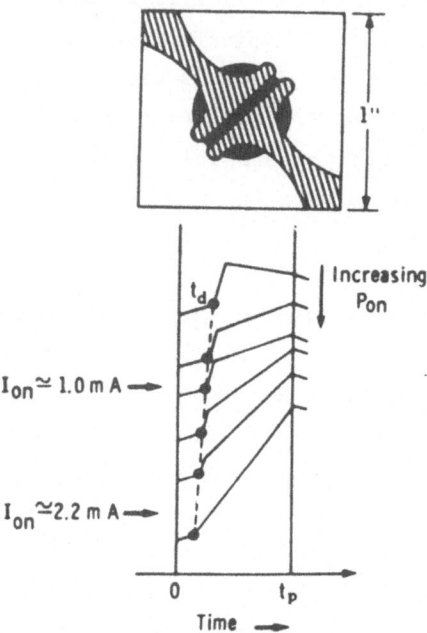

Fig. 5. Top: sample geometry viewing through the bottom of the 1-in.
glass substrate. The molybdenum electrodes are cross hatched
and the sample is black. The gap in the center, where
switching occurs, is exaggerated for clarity. The microphone
(thermophonic) cell is placed over the gap on the top of the
substrate. Bottom: typical microphone voltage response
curves for a series of single-shot threshold switching events
as a function of ON-state power; $t_p \cong 2$ ms. The microphone
signal is a property of the detector, we express it in arbitrary voltage units. The top trace was taken on a virgin
sample. Each successive trace corresponded to an increase in
ON-state power, P_{ON}, produced by reducing the load resistance
in the circuit.

was only after the ON-state power was reduced from its maximum value that a memory event occurred; the crystalline memory filament was clearly evident under optical and scanning electron microscopy.

The data shown in Fig. 5 were taken simultaneously with the voltage-time, V(t) and current-time, I(t), traces. Comparison of these traces with the thermophonic signal clearly showed that the switching event occurred well before the capacitive jump in the microphone signal (typcially 40 μsec earlier; the propagation time of the signal generated from the surface of the sample to the microphone was 6 μm). That is to say, the experiment convincingly demonstrated that an electronic event occurs prior to the occurrence of unusual thermal processes in the film.

To explain the observed signals in a quantitative manner and develop a qualitative model for the various phenomena, a modification of the Aamodt-Murphy[14] model was used to numerically simulate the results. The heat flow equation was solved in four regions (substrate, sample, gas, cell) with the temperature and its spatial derivative set equal at the three interfaces. The basic model was modified by introducing time-dependent step function thermal conductivities to model the transition in the physical properties of the sample and the resulting increased effectiveness of the electrodes as heat sinks. Since the electrodes could not be treated in the one-dimensional approach, their thermal properties were simulated via a change in the thermal conductivity of the substrate. The data were fit with two parameters, the substrate effect and the Joule heat power density H(t) developed in the film, and the expected microphone signal and temperature at both the sample-gas and sample-substrate interfaces was calculated assuming that the heating occurs uniformly in a narrow filamentary path both prior to and after switching. (The volume of this formed path was determined by scanning electron microscopy.) In the inset of Fig. 6 the H(t) employed to fit a particular representative threshold switching event is shown. H_1 represents the input power density developed during the delay time $(t < t_d)$, H_2 represents the capacitive discharge, and H_3 is the power density developed during the ON-state period. In Fig. 6 we see that the average temperature reached at the sample-gas interface at t_d is about 100°C. Note that when the voltage is below that required for switching, the experimental nonswitching signal continues to increase linearly with time (as in the switching signal with $t < t_d$). Indeed, this linear non-switching signal in what is expected from the Aamodt-Murphy model. More

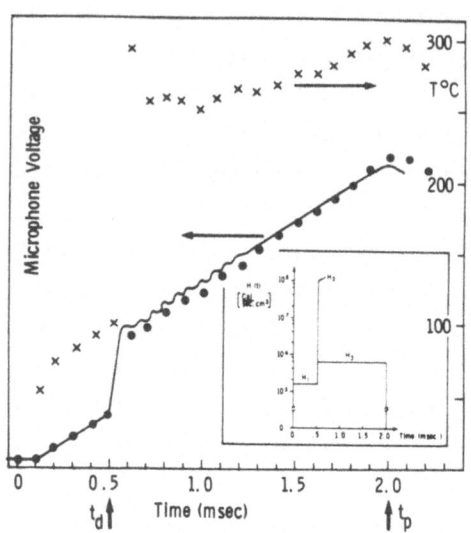

Fig. 6.   Microphone signal (solid-experimental, dots-calculated) and calculated temperature at the sample-gas interface (crosses) as a function of time for a threshold switching event. Calculations[7] suggest that the oscillations in the experimental signal may be due to a radial mode in the cell. The thermal conductivity of the substrate was increased by a factor of 2.3 at t_d; the thermal conductivity of the sample was increased by a factor of 5.3. The inset is the time-dependent power density used to fit the data. Because of the coarseness of the grid, we expect that the actual temperature reached during the capacitive discharge is higher than the calculated value.

importantly, no apparent differences in the nonswitching and pre-switching signals are observed.

In order to fit the threshold and/or memory switching data when the capacitive discharge was not mirrored in the microphone signal, which occurred for well formed samples, $H_2 = 0$ and $H_3 > H_1$ was chosen in a representative memory case that was quantitatively modeled. (Here the average temperature reached at t_d was only about 75°C) $H_2 = 0$ was taken on the following physical grounds. The capacitive discharge occurs in a few microseconds, whereas the pulse duration is some milliseconds. During the discharge the capacitive energy is initially given to the formed filamentary region, which acts as a heat source ($H > 0$) as it is rapidly heated to the vicinity of the melting temperature. However, just after this, the filament acts as a heat sink ($H < 0$); the heat of transformation required for melting is provided to the crystallites. The integrated $H(t)$ was assumed to vanish for the duration of the capacitive discharge and for a short time (~1 μs) thereafter during which the filament melts.

The capacitive discharge plays a major role in determining the extent of heating, melting, damage, and forming. The first phase of the discharge involves gross increases in conduction current while the voltage is held fairly constant by intrinsic and package inductive components[4]. The conduction current reaches values substantially higher than the threshold value and substantial Joule heating is expected. But, as the top trace in Fig. 5 shows, the discharge itself need not cause changes in the properties of the sample. Here, we see a negative slope in the post-switching signal, suggesting that cooling can occur from the discharge until the end of the applied pulse. (This particular event might involve some form of melting but with rapid cooling during the ON state so that complete vitrification occurs and forming is not observed.) Once the ON-state power is increased above a value where local phase separation and crystallization is encouraged, however, subsequent events involve some of the discharge energy going into melting of the crystalline part of the formed region rather than the heating of the filamentary path.

The ON-state operating point is crucial. If high power levels are involved, the filament continues to heat throughout the pulse duration.

The explanation for why one has to reduce the power level to obtain a memory event in Fig. 5 is that the temperatures reached in the filament at the end of the pulse exceeded the melting temperature of the locked ON-state crystalline path. The cessation of the pulse then caused rapid revitrification to the OFF-state. The sequence of events leading to a locked ON memory state is that sufficient forming has had to have occurred in a previous cycle so that the capacitive discharge can cause melting of the entire path between electrodes. If the temperature then reduces to the crystallization range ($240°C<T<340°C$ in the material studied), a highly conductive crystalline path will grow between the electrodes. If the temperature at the end of the pulse is kept below the melting temperature, T_M ($\cong 340°C$), a memory event will result. If the temperature exceeds T_M, remelting and then revitrification will then drive the sample back to the OFF state.

To explain the shortened t_d and the lower calculated temperature at t_d when forming occurs, it is suggested that the formed sample is effectively shortened, either by the dispersal of metallic crystallites and/or the extension of a crystallite from the electrode. In this case the temperature required for the attainment of a critical electric field near the electrode is reduced; t_d is therefore also reduced.

Finally, consider the microphone signal that occurs during t_d. The average temperatures do not approach those required to melt or perhaps phase separate the sample, although they may reach the glass transition temperature ($T_c \cong 140°C$). The linearity in the microphone signal during t_d, and the maintenance of its nonswitching form argues agains the development of thermal instabilities. It appears likely that the switching event is triggered by an electronic effect such as the attainment of $\mathscr{E}_c$ near an electrode. For short enough pulses in thin enough samples this is certainly the case[3,4]. For thick films this is a surprising result, but since the temperatures predicted during the discharge and ON state are in good quantitative agreement with what is required to cause changes in the properties of the sample, we think that the preswitching estimates of the temperature are reliable.

The switching and memory events outlined above concur with a model where the sample heats during the delay time; the heating causes the electric fields to rearrange and a critical value is reached near an electrode, whereby a carrier generation process is encouraged. During t_d, voltage reversals[15] or pulse bursts[16] will have no effect on t_d, since a critical local power density[4] seems to be required to drive the electric fields sufficiently high near the electrodes. Several models have been proposed in the past to establish the electronic process that causes the switching event to occur[2,3,4]. Thermophonic experiments do not seem to be able to determine which model is most appropriate. However, once the switching event occurs, the resulting effects on the morphology of the sample are determined by the capacitive discharge energy and the ON-state power level, as discussed above. For the particular material discussed the criteria for a memory event to occur in a well formed sample were simply that: (1) melting occurred at the switching transition; (2) the temperature after switching was in the crystallization range 240°C < T < 340°C; (3) the temperature at the end of the pulse remained below the melting point of about 340°C. Therefore, two distinct ranges exist where threshold switching events occur: low ON-state power, where phase changes are absent or minor; high ON-state power, where T_M is exceeded at the end of the pulse. Indeed, in the high power mode of electronic switching, the sample melts twice during a switching cycle, making it understandable why it was often suggested in the past that thermal effects were fundamental to the switching process. In thin-film sandwich samples, of higher glass transition temperatures and resistances, typical capacitive discharges may not provide sufficient energy for memory events to occur, and if care is taken not to overdrive these samples, the forming process is readily avoided[3,6,7].

I am grateful to Stanford Ovshinsky and many members of the staff at Energy Conversion Devices for their support and assistance throughout the past 17 years. I have also benefitted from discussions on switching with the late David Adler, John DeNeufville, Edward Fagen, Richard Flasck, Helmut Fritzsche, Irving Gastman, George Cheroff, Harold Grubin, Heinz Henisch, Scott Holmburg, Sergey Kostylev, James Kotz, Arun Madan, Simon Moss, Kurt Petersen, Howard Rockstad, Marvin Silver, and Peter Solomon.

References

1. S.R. Ovshinsky, Phys. Rev. Lett. __21__, 1450 (1968).

2. A. Madan and M.P. Shaw, The Physics and Applications of Amorphous Semiconductors, (Academic Press, N.Y., in press).

3. D. Adler, H.K. Henisch and N. Mott, Rev. Mod. Phys. __50__, 209 (1978).

4. M.P. Shaw, Physics of Disordered Materials, p. 793 (Plenum Press,

N.Y., 1985); D. Adler, M.S. Shur, M. Silver and S.R. Ovshinsky, J. Appl. Phy. 51, 3289 (1980).

5. H. Fritzsche and S.R. Ovshinsky; J. Non. Cryst. Solids 2, 393 (1970).

6. J. Kotz, Ph.D. Thesis, Wayne State University, 1982; J. Kotz and M.P. Shaw, Appl. Phys. Lett. 42, 199 (1983).

7. J. Kotz and M.P. Shaw, J. Appl. Phys. 55, 427 (1984).

8. M.P. Shaw and K.F. Subhani, Sol. St. Electronics 24, 233 (1981).

9. W.D. Buckley and S.H. Holmberg, Sol. St. Electronics 18, 127 (1975).

10. M.P. Shaw, S.H. Holmberg and S.A. Kostylev, Phys. Rev. Lett. 23, 521 (1973).

11. R.W. Pryor and H.K. Henisch, J. Non Cryst. Sol. 7, 181 (1972).

12. K.W. Petersen and D. Adler, J. Appl. Phys. 47, 256 (1976).

13. N.F. Mott, Contemp. Phys. 10, 125 (1969).

14. L.C. Aamodt and J.C. Murphy, J. Appl. Phys. 49, 3036 (1978).

15. I. Balberg, Appl. Phys. Lett. 16, 491 (1970).

16. D.K. Reinhard, Appl. Phys. Lett. 31, 527 (1977).

INTERPRETATION OF RECENT ON-STATE AND PREVIOUS NEGATIVE

CAPACITANCE DATA IN THRESHOLD CHALCOGENIDE AMORPHOUS SWITCH

G. C. Vezzoli*

U.S. Army Materials Technology Laboratory
Ceramics Research Division
Watertown, Massachusetts 02172-0001

M. A. Shoga

Hughes Aircraft Company
P. O. Box 9219 Airport Station
Los Angeles, California 90009

ABSTRACT

Recent measurements studying the on-regime of the transient on-state characteristics (TONC) of an amorphous semiconductor Ovonic threshold switch, employing precisely balanced circuitry and isolated device voltage determination, have shown that the blocked on-state develops after an interruption subholding-voltage-time of about 100 ns. If the voltage interruption time below the holding voltage (V_h) is no greater than approximately 65-95 ns, the blocked on-state will <u>not</u> develop and the I-V curve will display a metal-like behavior. This is in agreement with a recombination single injection model for threshold switching in amorphous semiconductors because during the interruption-time recombination takes place near the anode until a recombination front is established. In both the slow and fast relaxation regimes the general I-V curves thus derived are normally asymmetric about the origin, the former (slow) because of

*Also

Visiting Scientist
The Massachusetts Institute of Technology
Materials Science Department
Cambridge, Massachusetts 02138

and

Adjunct Professor
Department of Electrical Engineering
Rutgers University
Piscataway, New Jersey 08855-0909

the structure of the interrogating wave form, the latter (fast) because of the lag in the response of current to voltage. The present paper gives a conceptual band structure model to interpret the new I-V data based on a shallow trapping band, using a recombinative single injection mechanism to describe threshold switching. This model is also invoked to explain the previously-measured negative capacitance in this material in terms of carrier separation characteristic of a relaxation semiconductor undergoing threshold switching and involving a recombination front.

INTRODUCTION

The transient on-state (TONC)[1,2] of an amorphous semiconductor Ovonic threshold switch represents the I-V device characteristics which arise during the fast electrical deformation of the on-state. These "deformations" are imposed by the application of "rare" single and almost-square diagnostic pulses which cause the voltage to drop at a specific rate to specific values (including zero) for specific time intervals.[3,5] TONCs thus derived in the past showed symmetric I-V characteristics given in Fig. 1. The almost-horizontal regime of the I-V in Fig. 1 is referred to as the blocked on-state.

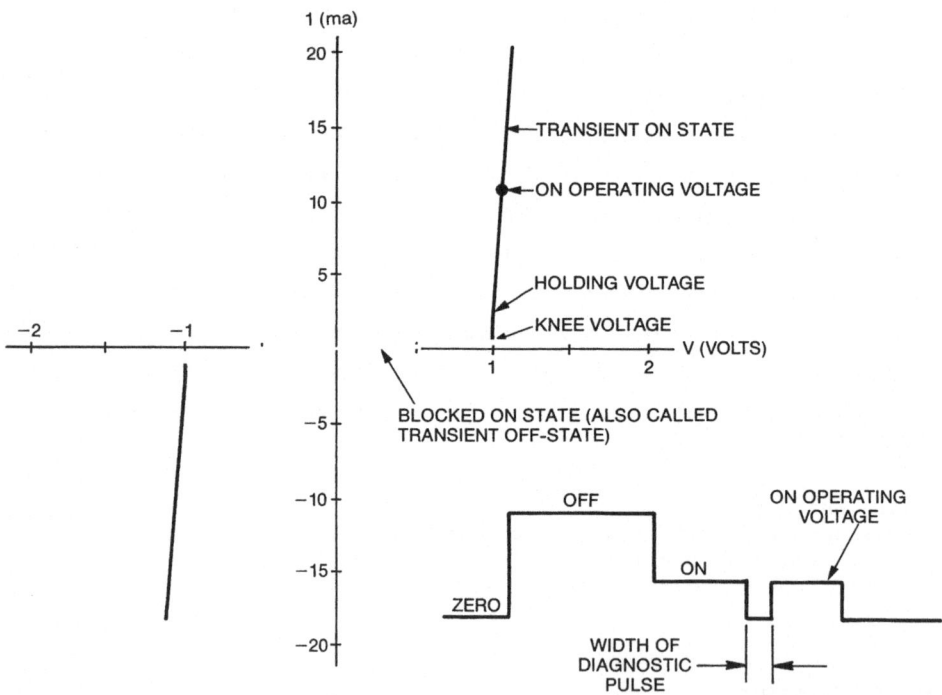

Fig. 1. The TONC I-V characteristics and the set-pulse followed by the operating on-voltage and a diagnostic interruptives probing pulse. The symmetric TONC is derived from the work of Ref. 1-5.

Recent measurements[6] using improved balanced and isolated circuits, and precise device voltage values, have shown that during very fast ramp interruption of the holding voltage to $V = 0$, and reimposition to $V = V_h$, the blocked on-state is by-passed or unestablished (monotonic curves in Fig. 2 (A1 and B1); circuit given in Fig. 3). The maximum allowed time beneath the holding voltage $[\tau\ (V < V_h)]$ is about 95 ns in order to preclude the development of the blocked on-state of the TONC. The asymmetry of this I-V curve, or the failure to pass through the origin is due to the time required for current decay at zero field. This time interval is the sum of the trap liberation and carrier scattering times. The non-monotonic curve shows that for $(V < V_h) > 100$ ns, the blocked on-state will develop. The value of 100 ns is taken as the time constant, τ_1, and is interpreted subsequently. The asymmetry in this case is due to the impressed wave form and the level of the voltage at τ_1. The experimental value of τ_1 of about 100 ns in this study compares to a value measured to be about 45 ns in the less precise study of previous work given in Fig. 4 from Ref. 4.

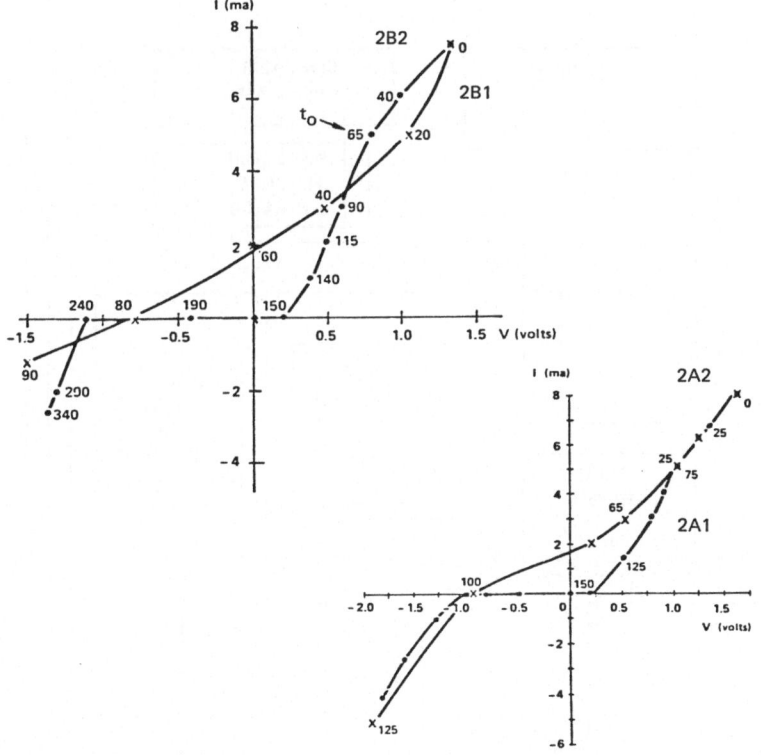

Fig. 2. The TONCs derived from Ref. 6 showing asymmetry regarding the development of the blocked on-state (horizontal subregime), and showing the bypassing of the blocked on-state occuring for very fast interruptions (Fig 2A1 and 2B1). These interruptions are of the order of less than 100 ns.

From comparison of Fig. 4 to Fig. 2, and correlation with our previous related work[7] on NbO_2 we attempt now to advance a working model to describe the transition from on-state to blocked on-state to off-state. We also seek to show correlation with previously measured negative capacitance at switching,[8] (see Fig. 5).

DISCUSSION

The asymmetry shown in Fig. 2, albeit puzzling at first, is reasonable in that it rules out back-to-back barriers[9] in series with negligible-resistance bulk material (such identical barriers being very rare indeed in nature). Furthermore previous work shows that: (1) the off-state condition $\tau_d > \tau_\ell$ attempts to reverse its inequality at switching[4,10] (where τ_d = dielectric relaxation time, and τ_ℓ = diffusion length life-time); and (2) electrophoto-luminescence intensity is polarity-dependent.[11] These observations taken together suggest a relaxation semiconductor material having a recombination front nearer to the anode than the cathode.[10] The relaxation semi-conductor threshold switch re-

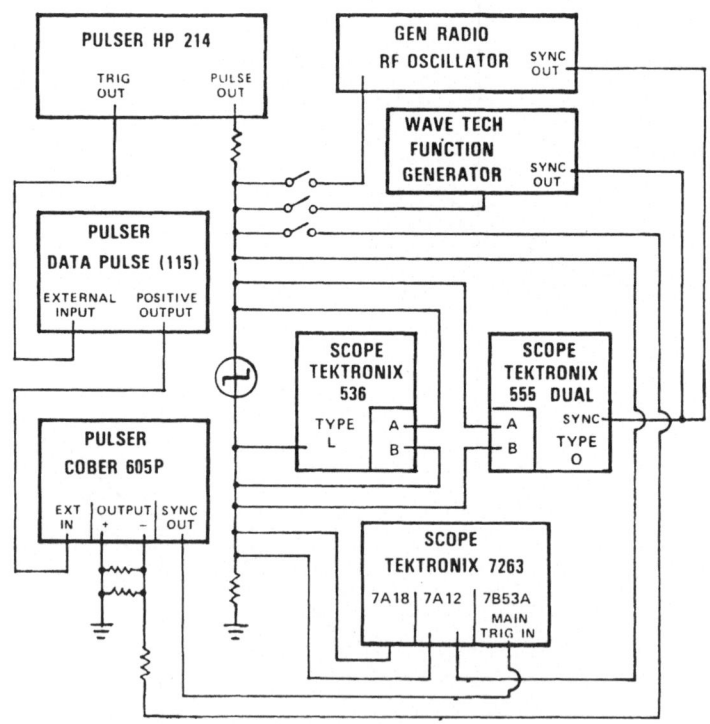

CIRCUIT SCHEMATIC FOR SWITCHING AND
DIAGNOSTIC PROBING DESCENDING RAMP

Fig. 3. The circuit schematic for the work of Ref. 6.

70

quires a trap saturation with equal and opposite flow densities of electrons and holes and a minimum conductivity condition. This may occur at elevated concentrations which further decay, and a large-signal recombination front may form at an intermediate stage. In either such case the mechanism for switching is referred to as single recombinative space charge injection.[10] As the relaxation semiconductor is addressed by an increasing electric field, τ_d decreases. However, as the value of τ_d is decreasing and approaching the value of τ_ℓ, although the transition to the lifetime state of a semiconductor ($\tau_\ell > \tau_d$) seems imminent, it does not in fact occur. This is because the condition $\tau_d = \tau_\ell$ is instead reached and carrier separation takes place. This causes the polarizability to become extremely large.

Mathematically, the divergence of polarizability is indicated by the difference term ($\tau_d - \tau_\ell$) appearing in the denominator within a bracket in the original relaxation semiconductor analysis given by Van Roosbroeck et al.[12] This divergence further supports recombinative single injection as it relates to pre-switching behavior. Admittedly, the relaxation semiconductor mathematical analysis cannot be solved analytically in closed form without approximation, and must in the future be solved precisely by numerical methods. However, the implication of the above $\tau_d = \tau_\ell$ condition should remain the same under significance of approxi-

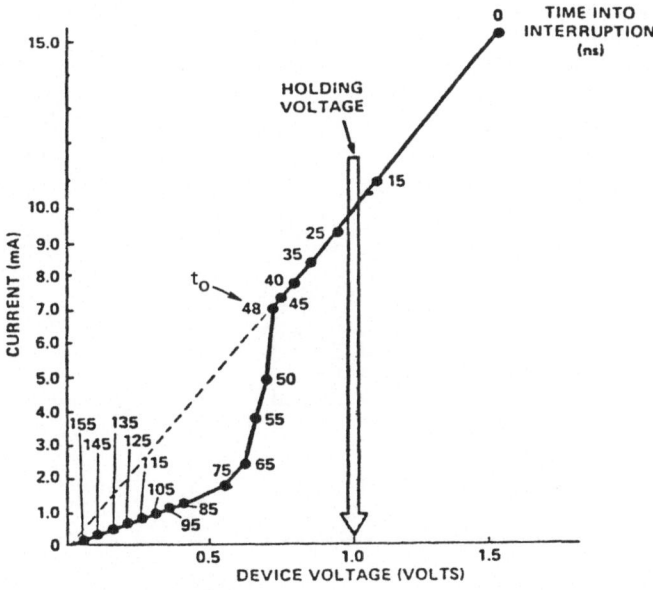

Fig. 4.   The I-V for the TONC from Refs. 3-5 derived in a point by point manner, and showing time into the interruption.

mation algebraic or precise numerical solution. (The exact signficance
of approximation in the former methodology is still the relaxation
semiconductor switching mechanism is absolutely necessary).[12,9]

In the transitional range in the neighborhood of $\tau_d = \tau_\ell$ the small
signal solutions involve oscillatory (rather than monotonic Bessel functions, and electrons and holes can move in opposite directions. This
qualitative behavior should occur also under large-signal conditions so
that when carriers are activated from traps during the pre-switching, the
transitional range is realized and the carriers are separated. Local
neutrality during pre-switching could be effected by the separation process.

Intuitively, it would seem surprising that the observation of negative capacitance at threshold conditions in Ovonic amorphous devices[8]
were not related to the mechanism of threshold switching. We suggest
that the phenomenon of carrier separation by which at threshold the
minority carrier is swept to the one electrode, and the majority carrier
is swept to the couterelectrode is the fundamental cause of the negative
capacitance singularity.[13] An analysis of this hypothesis by numerical
methods is currently underway by Henisch and Rahimi,[14] and by Callorotti
and Schmidt.[15]

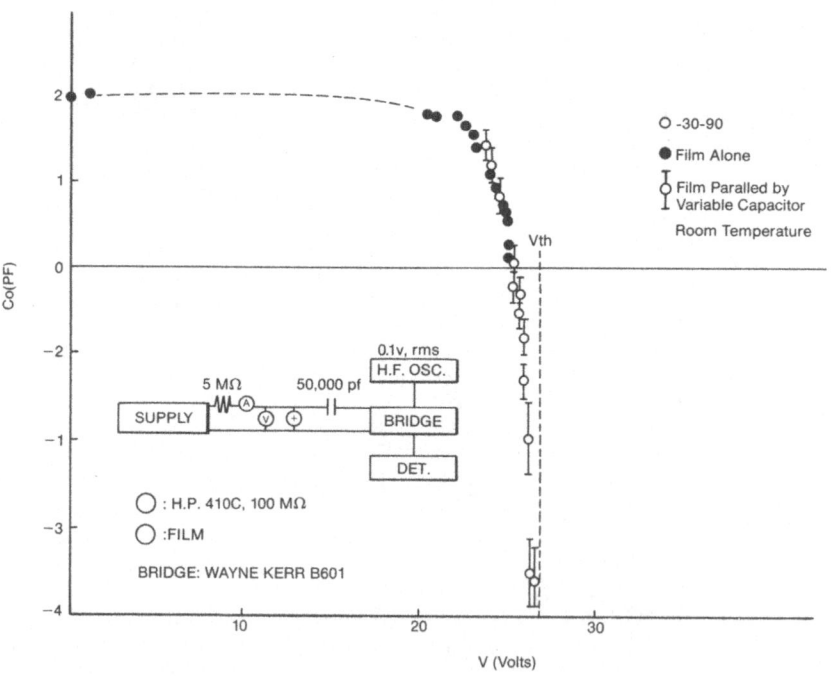

Fig. 5. The observation of negative capacitance near threshold conditions reproduced form Ref. 8.

We now trace the decreasing voltage cycle of Fig. 2 in an effort to develop a model the TONC based on shallow traps which govern the trap-limited recombinative space charge single injection.

As the voltage across the device is reduced below the operating voltage (about 1.4V in Fig. 2, curve B2) and then beneath the holding voltage (1V), the conduction band free carriers are de-energized and start to recombine into a shallow trapping band (schematized in Fig. 6). The recombination sites are available because trapped carriers are recombining in their own right into the valence band. This regime is shown in the I-V condition at $t = t_0 \approx 65$ ns (where the I vs V ungergoes major slope change). However, only the time interval between $t(V = V_h)$ and t_0 is significant because the physics at $V > V_h$ is due to overvoltage only (and creates hot electrons). This time is approximately 25 ns, agreeing reasonably well with the equivalent but somewhat less accurate data of Fig. 4 yielding 28 ns. At this time condition, t_0, the free carrier density or concentration is reduced to what we believe to be a critical

BAND STRUCTURE MODEL FOR TRAP-LIMITED SPACE CHARGE RECOMBINATIVE
SINGLE INJECTION THRESHOLD SWITCHING TO EXPLAIN TONC

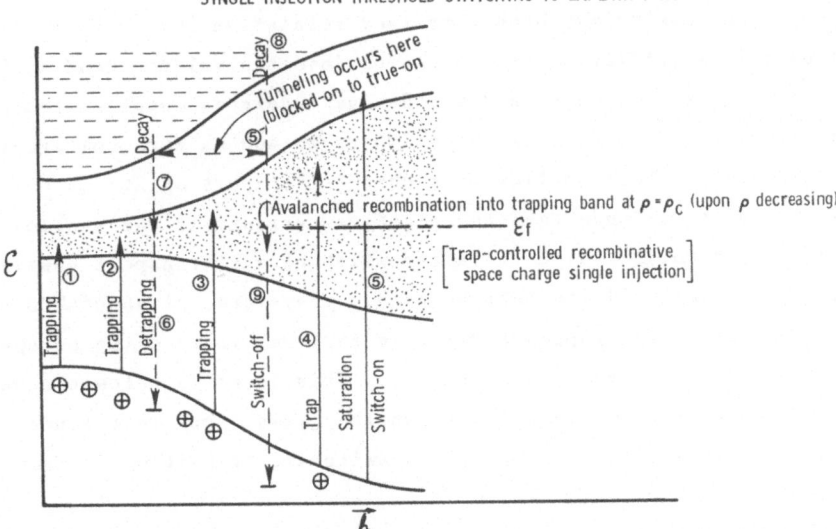

	(1-3)	Trapping occurs as field is raised. Conductivity is mildly p-type.
Switch-on 1-5'	(4)	At critical trapping level, or saturation, further injection occurs, trap-unimpeded, to cause populating of conduction band.
	(5)	The populating of the conduction band in an avalanche to cause switching. Carriers are n-type; electrons are space charge that are not compensated.
	(5)'	Tunneling mechanism for vertical regimes of TONC (transition from blocked-on state to true-on state).
Switch-off 6-9	(6)	Detrapping occurs into valence band.
	(7)	Decay occurs from conduction band into trapping band.
	(8)	Avalanched recombination at $\rho = \rho_c$.
	(9)	Switch-off from trapping band to valence band.

Fig. 6. Schematized band model to describe the TONC and the switching-off characteristics of an amorphous chalcogenide threshold device.

level ρ_c. Below ρ_c the free carrier concentration is <u>not</u> sufficient to screen the Coulombic attractive potential between electrons and holes (Mott Transition Type effect), and massive avalanche recombination then occurs into the trapping band. This occurs in Fig. 2b(2) during the region t≈65 ns to t≈150 ns. (The corresponding region in Fig. 4 occurs from t≈48 ns to t≈75 ns). During these 85 (or 10^2) ns in Fig. 2B(2) the conduction band empties and blocked on-state carrier transport occurs in the shallow trapping band. In order that the trapping band is not completely filled at this time, the trap band recombination to the valence band must proceed at the minimum at a slightly faster rate than free carrier recombination into the trapping band. This places an upper limit of about 100 ns for the value of the distribution trapped carrier lifetime which we refer to as the τ_1 alluded to earlier. Thus a critical number of traps must empty within the first 25 ns after the $V < V_h$ condition is established in order that ρ is allowed to reach ρ_c.

A reimposition of increasing voltage in the TONC during blocked on-state conditions allows the true on-state to be re-established. This re-establishment is hyposthesized to be due to tunneling (Fig. 6) because decay has not yet caused the level of trap population to fall below the bottom of the conduction band. Further relaxation at $V < V_h$ of the order of 1 us allows sufficient recombination into the valence band such that the trapping band is empty and the device undergoes total switch-off.[16] Partial switch-off occurs in about 200 ns to 400 ns zero-voltage-interruption time.[16] This implies that charge transport exists as positive carriers in the valence band and negative carriers in the shallow trapping band until 1 us decay after which charge transport continues only in the almost-filled valence band. Therefore, at about 200 ns to 400 ns of decay the trapped carrier release has reduced the occupancy of the trap band to a level less than E_f. This level is below a conduction band minimum (which we know exists in the glass) such that tunneling is impossible and at least a partial re-switching transition is inevitable.

ACKNOWLEDGEMENTS

The authors are indebted to W. Van Roosbroeck for reading the manuscript and proposing suggestions for clarifying the ideas advanced herein.

REFERENCES

1) R. W. Pryor and H. K. Henisch, J. Non. Cryst. Solids <u>7</u>, 1181 (1972).

2) H. K. Henisch, R. W. Pryor, and G. J. Vendura, J. Non-Cryst. Solids <u>8-10</u>, 415 (1972).

3) G. C. Vezzoli and L. W. Doremus, J. Appl. Phys. 44, 3245 (1973).

4) G. C. Vezzoli, Phys. Rev. B 22(4), 2025 (1980).

5) G. C. Vezzoli, Phys. Rev. B 22(4), 2025 (1980).

6) G. C. Vezzoli, Phys. Rev. June 1987.

7) G. C. Vezzoli, L. W. Doremus, M. Shoga, B. Lalevic, and S. Levy, J. Appl. Phys.

8) P. J. Walsh and R. Vogel, Appl. Phys. Lett. 14, 216 (1969).

9) H. K. Henisch, The Pennsylvania State University, private communication.

10) W. Van Roosbroeck, J. Non-Cryst. Solids 12, 232 (1973); Phys. Rev. Lett. 28, 1120 (1972).

11) G. C. Vezzoli, L. W. Doremus, P. J. Walsh, P. J. Kisatsky, J. Appl. Phys. 45, 4534 (1974).

12) W. Van Roosbroeck, Phys. Rev. 123(2), 474 (1961).

13) The charge separation is unique to a relaxation semi-conductor such as the Ovonic glass and GaAs, and definitely does not occur in a lifetime semiconductor. Hence we do not expect negative capacitance in the latter.

14) H. K. Henisch, The Pennsylvania State University and Saeid Rahimi, Sonoma State University, private communications.

15) P. Schmidt and R Callorotti, private communication.

16) G. C. Vezzoli, L. W. Doremus, and P. J. Walsh in "Amorphous and Liquid Semiconductors," edited by J. Stuke and W. Brenig (Taylor and Francis, London, 1974), Vol. I, P. 651.

CARRIER INJECTION INTO LOW LIFETIME (RELAXATION) SEMICONDUCTORS

J.C. Manifacier, Y. Moreau, and R. Ardebili

Centre d'Electronique de Montpellier
Universite des Sciences et Techniques du Languedoc
-34060- Montpellier Cedex (France)

ABSTRACT

The paper deals with the consequences of a steady state minority carrier injection through a metal or a high-low junction into the bulk of a semiconductor. Depending on the nature of the semiconductor, the spatial distribution of the net recombination rate of injected minority carriers R occurs in two different ways:

(i) When the lifetime, τ_O , of the excess electron-hole pairs is much higher than the dielectric relaxation time τ_D, space charge vanishes in a distance on the order of the screening length L_s. L_s reduces to the Debye length L_D if trapped space charge can be neglected. The injected minority carriers and the neutralizing majority carriers decrease then with a characteristic ambipolar diffusion length $L_{Da} \gg L_s$ and so does the rate of recombination R.

(ii) A second class of semiconductors (relaxation semiconductors) are characterized by $L_s \gg L_{Da}$. They respond to injection in an entirely different way. The recombination rate is now highly localized inside a recombination front whose extension is of the order of L_{Da} and whose position is current dependent. Results obtained by numerical simulation and analytical modeling are presented.

INTRODUCTION

When dealing with charge transport into semiconductor devices, we must solve some phenomenological transport equations adapted to the physical problem under study.

For the present two carrier system, the complete equations to be solved are for the steady state and assuming Boltzmann's statistic.[1,2]

1) The current equations, for a one-dimensional geometry:

$$J_n = e \, \mu_n \, n \, E + \mu_n \, kT \frac{dn}{dx} \qquad (1)$$

$$J_p = e \, \mu_p \, p \, E + \mu_p \, kT \frac{dp}{dx} \qquad (2)$$

The carrier mobilities, μ_n and μ_p , will be taken here as constants though when searching for a numerical solution it is a trivial matter to consider position or field dependent mobilities.

2) Poisson's equation:

$$\frac{dE}{dx} = \frac{e}{\epsilon} \, [\, p - n + N_D - N_A + Q_r \,] \qquad (3)$$

We have assumed complete ionization of donors and acceptors. Qr is the space charge localized in the traps or recombination centers. If we assume a center with two states of occupancy, neutral when occupied and positively charge when empty, we can write:

$$Q_r = p_r \qquad (4)$$

where p_r is the density of empty centers. For a

recombination center density N_r, we have:

$$N_r = n_r + p_r \tag{5}$$

The recombination center occupancy is a function of the free carrier densities in the conduction and valence bands as well as of the trap parameters.

In the steady state and for Shockley-Read recombination statistic [3] the density of empty centers is given by:

$$p_r = N_r \frac{\tau_{ne} p + \tau_{pe} n_1}{\tau_{pe} (n + n_1) + \tau_{ne} (p + p_1)} \tag{6}$$

3) The particle continuity equations:

$$+ \frac{1}{e} \frac{dJ_n}{dx} + \text{Generation} - \text{Recombination} = 0 \tag{7}$$

$$- \frac{1}{e} \frac{dJ_p}{dx} + \text{Generation} - \text{Recombination} = 0 \tag{8}$$

If we assume thermal generation only, we can write:

$$\text{Recombination} - \text{Generation} = R = \frac{np - n_i^2}{\tau_{pe} (n + n_i) + \tau_{ne} (p + p_i)} \tag{9}$$

with $n_e p_e = n_1 p_1 = n_i^2$.

These transport equations can be written as three coupled differential equations involving the electric field E (or electric potential Ψ ($E = -\frac{d\Psi}{dx}$)) and the free carrier concentrations n and p. We will present here some solutions obtained for the case of low lifetime semiconductors (relaxation semiconductors).[4-6]

The complete set can be solved in three different ways:

(i) using numerical modeling; [7-10]

(ii) by a linearization procedure around a steady state value of polarization [11-13] ('small signal analysis'); and

(iii) by using some simplifying assumptions, one can obtain analytical solutions for a given range of polarization (or currents).[14,15]

In the following sections, we will discuss some numerical and analytical results obtained for P^+-N as well as P^+-ν-N^+ structures. Before that, we must recall the criterion given by Van Roosbroeck [4,5] which classifies the semiconductors according to their behavior under free carrier injection. This behavior is drastically different for low lifetime semiconductor as compared to the familiar high lifetime semiconductor (so called lifetime semiconductor) such as silicon or germanium.

LIFETIME AND RELAXATION SEMICONDUCTORS

A large amount of research publications deals with the transport of charge carriers in various semiconductor devices. When looking for an analytical solution, for example the current-voltage relationship of a given structure, it is usually correct to assume space charge neutrality into the homogeneous bulk far from space charge region at the contacts.

This simplifying hypothesis is often valid for such semiconductors as silicon or germanium at room temperature. Space charge neutrality condition arises from the much lower dielectric relaxation time $\tau_D = \epsilon / \sigma_e$, where σ_e is the equilibrium conductivity, as compared to the recombination time τ_O characteristic of the recombination center. For a transient phenomenon, the dielectric relaxation time is closely related to the time necessary for the space charge to disappear. In the realm of a homogeneous (no diffusion) small space charge departure from equilibrium, the neutrality is restored with the following time dependence:

$$(\Delta p - \Delta n + \Delta Q_r) \sim \exp(- \frac{t}{\tau_D}) \qquad (10)$$

The recombination time, on the other hand, measures the recombination of excess electron-hole pairs. Following the Shockley-Read scheme, the diffusion length lifetime τ_0 is given by:[5,7]

$$\tau_0 (n_e + p_e) = \tau_{pe} (n_e + n_1) + \tau_{ne} (p_e + p_1) \qquad (11)$$

For a low injection level and assuming space charge neutrality and no traps: $\Delta n \approx \Delta p$

$$R \approx \frac{\Delta n}{\tau_0} \approx \frac{\Delta p}{\tau_0} \qquad (12)$$

The above defintion is valid for small departures from equilibrium. For a high level of carrier injection, these characteristic times τ_0 and τ_D lose their precise meaning, but they are still useful as shorthand symbols to characterize a given semiconductor.

Van Roosbroeck was the first to point out the difference of behavior [4,5] under minority carrrier injection between lifetime semiconductors ($\tau_D \ll \tau_0$) and relaxation semiconductors ($\tau_D \gg \tau_0$).

Under nonequilibrium, in high "lifetime semiconductors," minority carrier injection will induce an increase in the majority carrier concentration in order to diminish the net space charge density. In the "relaxation semiconductors," on the other hand, injected minority carriers will recombine, and this in turn will induce a space charge build-up:

$$\Delta p = p(x,j) - p_e(x,0) > 0$$

and $\qquad \Delta n = n(x,j) - n_e(x,0) < 0$

We will deal here with steady state conduction only. In that case, characteristic lengths are more adapted to the interpretation of the physical behavior than characteristic times. The characteristic length necessary for the free carriers to screen a fixed space charge is the Debye Length L_D:

$$L_D = \left[\frac{\epsilon \, kT}{e^2 \, (n_e + p_e)} \right]^{\frac{1}{2}}$$

(13)

In the presence of traps the pertinent length is the screening length, L_s. For a trap density, N_r, much greater than the equilibrium majority carrier concentration n_e we obtain:

$$L_s \approx L_D \, [1 + N_r/n_e]^{-\frac{1}{2}}$$

(14)

L_s can be much shorter than L_D. Space charge exists at the boundary between different semiconductors too. The width W of the space charge layer for an asymmetric P^+-N or N^+-P junction is given by:

$$W \approx \left[\frac{2 \, \epsilon \, V_D}{e \, N_{D,A}} \right]$$

(15)

where V_D is the diffusion potential of the junction. At room temperature and for lifetime semiconductors such as germanium or silicon ($V_D \sim 20 - 40$ kT/e) and these characteristic lengths, L_D, L_s and W, where space charge effects prevail, have values on the order or lower than $1\,\mu$m.

In the steady state regime, the characteristic length associated with the recombination of excess carriers in-jected through a contact is the ambipolar diffusion length L_{Da}. For extrinsic semiconductors L_{Da} reduces to the minority carrier diffusion length. Typical L_{Da} values for Silicon are in excess of 10 μm up to hundreds of μm . It can be easily shown that for lifetime or relaxation semiconductors, we have [16]:

$$\frac{L_D}{L_{Da}} \approx \left[\frac{\tau_D}{\tau_o} \right]^{\frac{1}{2}}$$

(16)

The Van Roosbroeck criterion, based on the value of the characteristic times ratio, applies to the characteristic lengths ratio as well. When space charge in traps is taken into account, the relevant quantity is L_s/L_{Da}, and we will use the following classification:

$$\frac{L_s}{L_{Da}} < 1 \text{ corresponds to lifetime semiconductors.}$$

$$\frac{L_s}{L_{Da}} > 1 \text{ corresponds to relaxation semiconductors.}$$

The parameters τ_o, τ_D, L_{Da} and L_D (or L_s) can be easily calculated for a known semiconductor.

They lose their precise meaning at high injection levels (high currents) but are, anyway, very useful to predict the behavior of a given semiconductor. A given semiconductor can work either in a lifetime or in a relaxation regime simply by increasing the capture cross section of the recombination center (i.e. by decreasing the values of τ_{ne}, τ_{pe} and then τ_o). This will affect the value of the L_s/L_{Da} ratio without modifying the electric field and carrier concentration contours inside the structure at equilibrium (J=0). These two classes of semiconductors behave differently only under nonequilibrium condition.

We will present now some qualitative results for an injecting P^+-N homojunction. The N part of the structure will be assumed long enough to sustain ohmic conduction deep inside the bulk (semi-infinite sample). We leave to the next paragraph the presentation of the quantitative result for the P^+ - N and P^+ - ν - N^+ structures.

Figure 1 gives a comparative schema concerning the behavior of lifetime and relaxation semiconductors under minority carrier injection. The N zone is weakly doped and can be either a lifetime ($L_s \ll L_{Da}$) or a relaxation ($L_s \gg L_{Da}$) semiconductor. The P^+ zone is a hole injecting contact, the free carrier densities are taken to be constant at x = 0 and the P^+

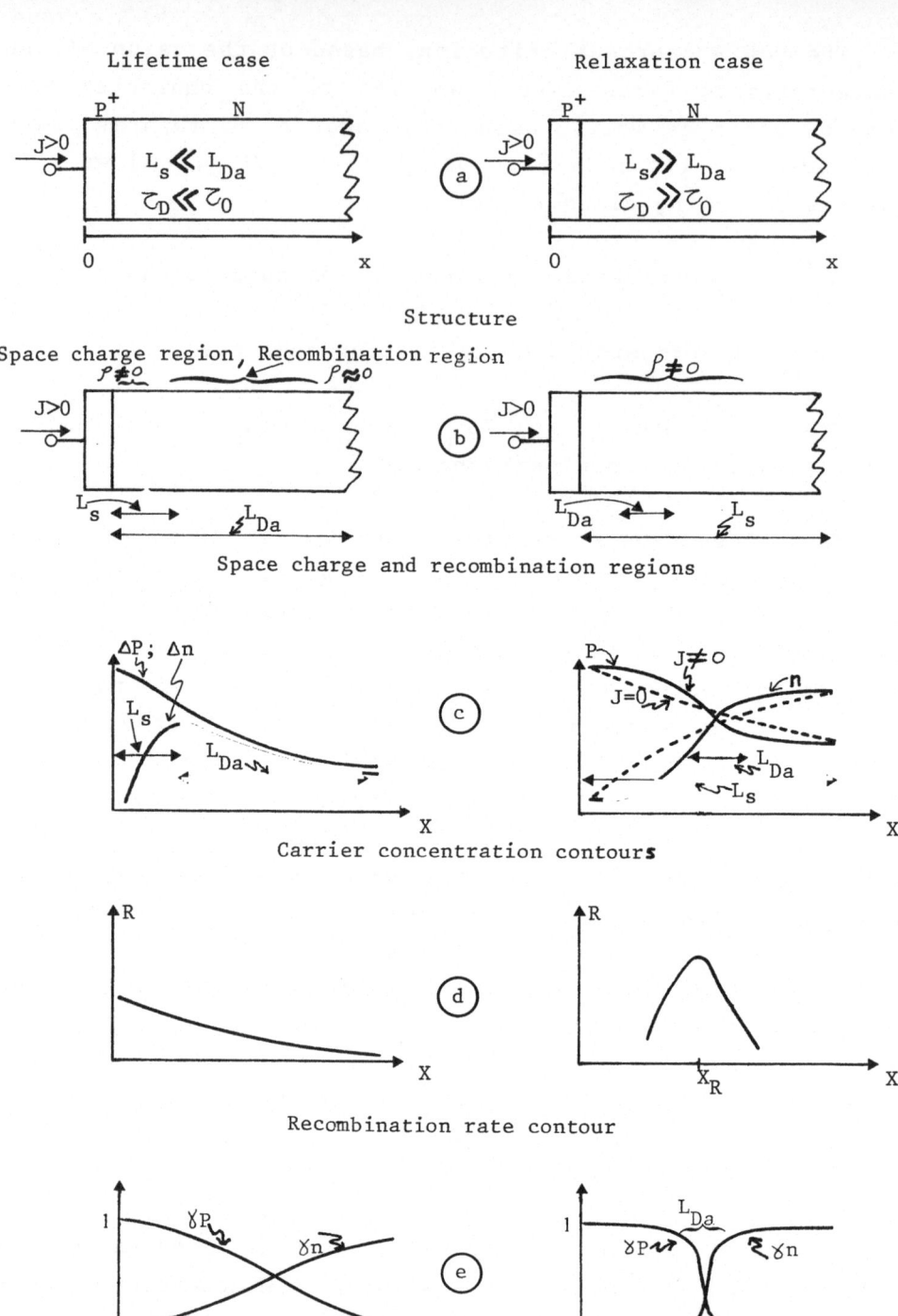

Fig. 1 Comparative behavior of a lifetime and a relaxation
semi-conductor.

zone is of a lifetime type. See the physical parameters in
Fig. 2.

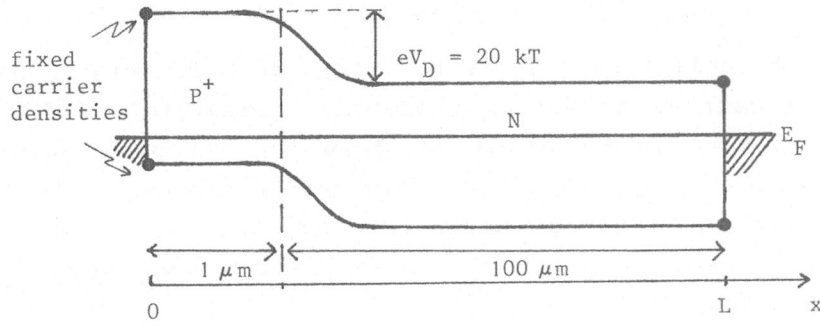

$$\epsilon_r = 11 : T = 300 \text{ K} : n_i = 3 \ 10^8 \text{cm}^{-3}$$
$$\mu_n = 8400 \text{ cm}^2/\text{V.s.} ; \mu_p = 400 \text{ cm}^2/\text{V.s}$$

P^+ zone:

$n_e = 3 \ 10^2 \text{cm}^{-3}; \ p_e = 3 \ 10^{14} \text{cm}^{-3}$

$N_t = 0$

$\tau_{ne} = \tau_{pe} = 10^{-8} \text{s.}$

$n_1 = 3 \ 10^9 \text{cm}^{-3}; \ p_1 = 3 \ 10^7 \text{cm}^{-3}$

$L_D (\approx L_s) = .229 \ \mu\text{m}$

$\tau_0 \approx 10^{-8} \text{s.}$

$L_{Da} \approx 14.74 \ \mu\text{m}$

N zone:

$n_e = 1.5 \ 10^{11} \text{cm}^{-3}; \ p_e = 6 \ 10^5 \text{cm}^{-3}$

$N_t = 1.5 \ 10^9 \text{cm}^{-3}$

$\tau_{ne} = \tau_{pe} = 10^{-11} \text{s.}$

$n_1 = 3 \ 10^9 \text{cm}^{-3}; \ p_1 = 3 \ 10^7 \text{cm}^{-3}$

$L_D (\approx L_s) = 10.23 \ \mu\text{m}$

$\tau_0 \approx 1.02 \ 10^{-11} \text{s.}$

$L_{Da} \approx .103 \ \mu\text{m}$

Fig. 2   Configuration used in the present computation, energy
band diagram and physical parameters.

Figure 1-b shows the extension of the space charge and
recombination region inside the bulk. For a lifetime
semiconductor, space charge neutrality is reached after a
few screening length L_s from the P^+ contact:

$$\Delta p - \Delta n + \Delta Q_r \approx 0 \tag{17}$$

The excess electron-hole pairs recombine then in a few
L_{Da} until the equilibrium densities are reached (see Fig.
1-c where we have assumed $\Delta Q_r = 0$ for simplicity). When
the current J crossing the structure is increased, the injected hole carriers and the neutralizing electrons extend

deeper into the N zone. The recombination rate, equation (9), decreases monotonically from the P^+ injecting contact (see Fig. 1-d).

In a relaxation semiconductor, on the other hand, minority carrier injection leads to a completely different behavior. A space charge of free and trapped carriers exists in a length of the order of a few L_s (see Fig. 1-b). The space charge neutrality hypothesis is not valid anymore. The free carrier concentrations obey the equilibrium relation:

$$n\,p \approx n_i^2 \qquad (18)$$

except in a length of the order of a few ambipolar diffusion length L_{Da} ($L_{Da} \ll L_s$) in which the recombination rate is very high. On the left of this recombination front, the total current J is a hole current, with a hole injection ratio:

$$\gamma_p = \frac{J_p}{J} \approx 1 \qquad (19)$$

Minority carrier injection is associated in this region with "majority" carrier depletion (see Fig. 1-c), and that depletion in turn explains the low recombination rate. Far from the P^+ injecting contact, the current is an electron current: $\gamma_n = \frac{J_n}{J} \approx 1$, few injected holes have survived and again the recombination rate is low. In between, there is a very fast variation of γ (see Fig. 1-e) associated with an intense recombination rate (see Fig. 1-d).

One of the most interesting results; following minority carrier injection in a relaxation semiconductor, is due to the fact that the position of this recombination front is current dependent. This dependence is a complicated function of semiconductor parameters, in particular the mobility ratio (μ_n / μ_p) and the boundary conditions of the structure under study. We must note too that, contrary to the lifetime case for which the excess carriers recombine in a zone which is usually neutral at $J = 0$ (n_e, p_e

independent of x), the recombination front in a relaxation semiconductor occurs in the space charge region (n_e and p_e are position dependent). This, in turn, makes any simple interpretation more difficult.

NUMERICAL RESULTS

The numerical results given here correspond to a P^+ - N structure. The complete set of parameters is given on Figure 2. We assume fixed carrier concentrations at the two contacts: x = 0 and x = L. These concentrations correspond to the unperturbed material (flat band conditions).

The coupled differential transport equations are written in term of the three independent variables n(x), p(x) and $\Psi(x)$. Their values are computed at each point x_i of a mesh ($0 \leq i \leq 100$). For lifetime semicon- ductors, n(x) and p(x) can be computed at each mesh point independently. On the other hand, for relaxation semiconductors, there is a high coupling due to the recombination term in continuity equations, a coupled resolution [18-21] is necessary. A large system of equations is solved at each iteration giving a correction term for each variable at each point. The current discretization using Gummel's formula [17] have been found efficient, and the convergence is rapidly obtained except for high currents.

Figure 3 gives the value of the recombination rate R as a function of x for different values of the current crossing the structure. The recombination is highly localized in space and the position of the recombination maximum is function of the current J. This behavior is typical of a relaxation semiconductor as opposed to a lifetime semiconductor for which the recombination rate decreases monotonically with increasing values of x from the injecting contact. When the current increases, in our P^+ - N structure, the maximum of the recombination rate

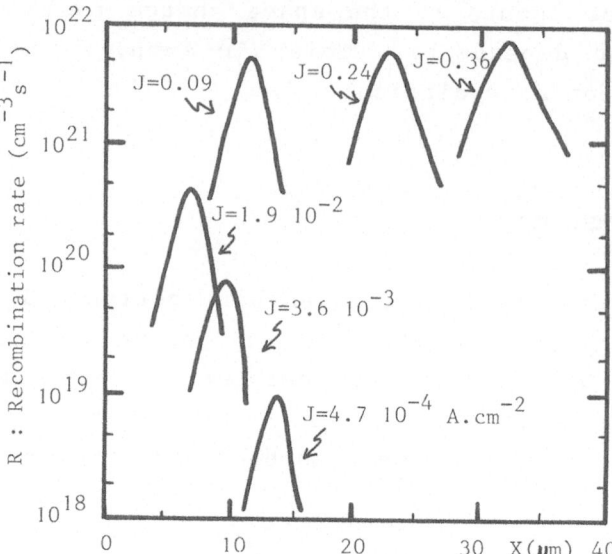

Fig.3 Net Recombination rate contour
      for a typical relaxation semi-
      conductor. The distance X is
      measured from the P⁺-N interface.

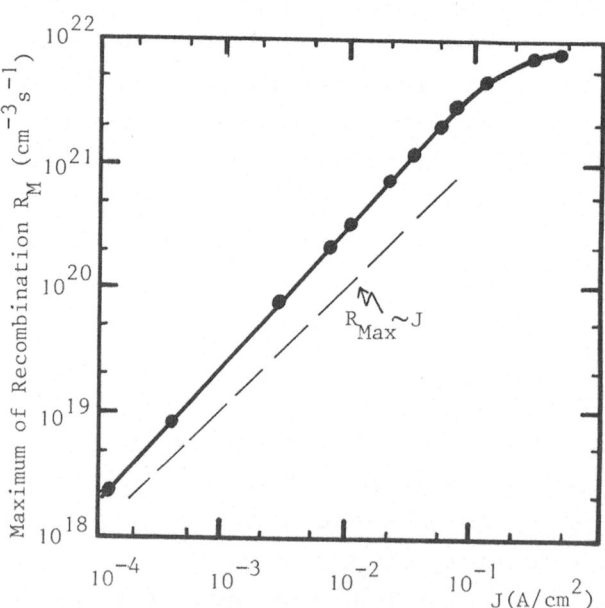

Fig.4 Value of the Maximum recombination
      rate as a function of current
      density for a typical relaxation
      semi-conductor.

first moves toward the injecting contact and then away at higher current densities.

Figure 4 shows the value of the recombination maximum R_{max} as a function of the current. The dependence is slightly supralinear with a trend toward saturation at high currents. R_{max} is much higher for a given current in the present case than in the lifetime case. From the continuity equations (7) and (8), and using the fact that across the recombination front the current injection ratio varies from one to zero, we obtain:

$$J = \int_{\text{recombination front}} e\, R(x)\, dx \tag{20}$$

The quasi linearity observed between R_{max} and J (see Fig. 4) allows us to write:

$$J \sim e\, R_{max}\, \Delta l \tag{21}$$

with l being a measure of the recombination front width whose value is practically constant at low currents, (see Fig. 3) and of the order of $1-2\,\mu m$. At high current, the recombination front broadening is associated with the observed saturation of R_{max} (see Fig. 4).

Figure 5 gives the current-voltage characteristic of the structure. In the range of the currents under study, we observe a linear variation at low voltage (ohmic part), followed by another linear variation at higher applied voltage.

DISCUSSION

At low current level, the conduction is contact controlled. The dominant mechanism is a recombination current inside the space charge of the P^+ - N junction.

At high current level, the analysis is best made by dividing the N part of the structure into three zones. The first zone corresponds to the region close to the P^+

contact; in this region the hole injection prevails ($J_n \approx 0$ and $\gamma_p \approx 1$). The current is a one carrier (holes) space charge current.[22] On the right side of the N region, the current is ohmic. This is imposed by the flat band condition at the right contact. In between, there exists a thin region where the hole space charge current adjusts to the electron ohmic current. This zone is the recombination front where the diffusion currents are intense and the voltage drop is limited to a few kT/e.

(a) high current level, P^+-N structure

We will neglect the voltage drop across the recombination front. Space charge injection of holes on the left will match with ohmic current on the right. Neglecting the voltage drop across the highly doped P^+ zone, we take the origin of the potential at the P^+ - N interface. V_a is the total applied voltage and $V(x_R)$ is the voltage at the recombination front (see Fig. 1-d and 1-e).

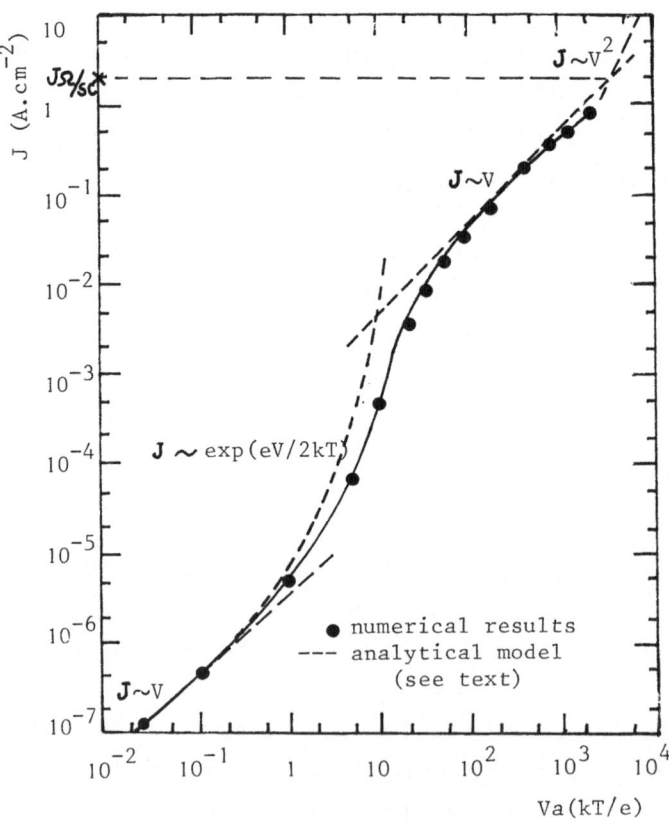

Fig. 5 Current-Voltage relationship for the P^+-N relaxation semiconductor structure.

The current voltage relationship for one carrier space charge current, the Mott–Gurney law, is obtained neglecting diffusion and the equilibrium carrier concentration:[22]

$$J = J_p = \frac{9}{8} \, \epsilon \, \mu_p \, \frac{[V(x_R)]^2}{x_R^3} \tag{22}$$

The hole concentration and electric field values at $x = x_R$ are given by:

$$p(x_R) = \left[\frac{\epsilon J}{2 \, e^2 \, \mu_p \, x_R} \right]^{\frac{1}{2}} \tag{23}$$

$$E(x_R) = \left[\frac{2 \, J \, x_R}{\mu_p \, \epsilon} \right] \tag{24}$$

(These equations are obtained as solutions of $J_p = e \, \mu_p \, p \, E$ and $dE/dx = e \, p/\epsilon$ with $E=0$ at $x = 0$. At $x = 0$, the neglected diffusion current will in fact satisfy current continuity.)

The condition of current continuity and electric field ar $x=x_R$ leads to:

$$e \, \mu_n \, N_D = \sigma_e \approx e \, \mu_p \, p(x_R) \tag{25}$$

and then to:

$$x_R = \frac{\epsilon}{2} \, \frac{\mu_p}{\sigma_e^2} \, J \tag{26}$$

At high current density, the recombination front moves away from injecting P^+ contact when J increases.

When $x_R \ll L$, the conduction is practically ohmic along the whole N region and the J–V relation is given by:

$$J \approx \sigma_e \, \frac{V_a}{L} \tag{27}$$

91

On the other hand, when the recombination front reaches the contact on the right side, the conduction is space charge controlled throughout the sample and J–V relation is given by:

$$J \approx \frac{9}{8} \epsilon \mu_p \frac{V_a^2}{L^3} \tag{28}$$

The cross–over between these two laws is obtained when x_R = L and, using the numerical parameters given on Fig. 2, the corresponding current value $J\Omega$ /so is found equal to 2.1 A/cm^2. These approximate analytical expressions for the J–V_a relationship have been plotted on Fig. 5. The difficulties encountered with the numerical resolution have limited the comparisons to current level under 1 A/cm^2.

(b) low current level, P$^+$-N structure

At very low current level, the recombination current inside the P$^+$-N junction is the controlling mechanism of conduction.[1] In the present case, for τ_{ne} = τ_{pe} we obtain:

$$J \sim e \, R_{Max} \, \Delta l \sim e \, \Delta l \, \frac{n_i^2 \, [\exp(eV_a/kT)-1]}{\tau_{ne} \, [n_1 + p_1 + 2 \, n_i \, \exp(eV_a/2kT)]} \tag{29}$$

Taking $\Delta l \approx L_{Da}$, this relation gives again a good fit with the numerical results (see Fig. 5). Numerical results have been obtained for other structures with injecting contact. The results obtained are explained using the same analysis.

(c) P$^+$-v-N$^+$ structure

This structure as well as the M-v-M structure are even simpler to analyze at high current level. Assuming one contact is hole injecting and the other electron injecting, we can divide the structure into two regions: a hole space charge current will match an electron space charge current at x_R, the position of the recombination front. This time a practically constant value is obtained for x_R:

$$x_R = \frac{\mu_p}{\mu_n + \mu_p} L \qquad (30)$$

and the $J - V_a$ relationship is given by:

$$J \approx \frac{9}{8} \epsilon (\mu_n + \mu_p) \frac{V_a^2}{L^3} \qquad (31)$$

This simple relation gives again a fairly good fit with numerical results published in the literature concerning relaxation semiconductors.[8,20]

One of the interesting results obtained above is the variation of the position of the recombination front with the current density for a P^+-N structure. If the recombination is radiative, a possibility then exists of having a spectral emission function of the current crossing the sample. This can be achieved either by varying the nature and physical properties of the radiative recombination centers through the structure, or else by considering a band to band radiative recombination across a variable energy band-gap.

To conclude, we have to point out the somewhat idealization for the structures analyzed above. In particular, for relaxation semiconductors corresponding to our N zone, the low lifetime value is a consequence of large cross section values (small τ_{ne} and τ_{pe}) and not of the high trap concentration. In practice, most semiconductors of low conductivity are highly compensated and, in turn, their high concentration of centers in the forbidden gap is associated with short screening lengths, L_s, making the semiconductor more lifetime like.

ACKNOWLEDGMENTS

Two of us, J.C. Manifacier and Y. Moreau, have had the privilege of working with Professor H.K. Henisch. Without him, this work could not have been done. Thanks are also due to Energy Conversion Devices, Inc. of Troy, Michigan for its continuous support.

REFERENCES

1. S.M. Sze, "Physics of Semiconductor Devices," 2nd edition, J. Wiley Interscience, New York (1981).

2. H.K. Henisch, "Semiconductor Contacts," Clarendon Press, Oxford, (1984).

3. W. Shockley and W.T. Read, "Statistics of the Recombination of Holes and Electrons," Phys.Rev. 87:835 (1952).

4. W. Van Roosbroeck and J.C. Casey, "Transport in relaxation semiconductors," Phys.Rev.B, 5:2154, (1972).

5. W. Van Roosbroeck, "Current-Carrier Transport with Space Charge in Semiconductors," Phys.Rev. 123:474 (1961).

6. F. Stockmann, "On the Concept of 'Lifetimes' in Photoconductors," R.C.A. Review 36:499 (1975).

7. C. Popescu and H.K. Henisch," Minority Carrier Injection into Semi-Insulators," Phys. Rev. B. 14:517 (1976).

8. G. Heder and O. Madelung, "The Influence of the Dielectric Relaxation on the Current Flow in High Ohmic Semiconducting diodes," Phys. Stat. Sol.(a), 30:215 (1975).

9. Y. Moreau, J.C. Manifacier and H.K. Henisch, "Minority Carrier Injection into Relaxation Semiconductors," J. Appl. Phys. 60:2904 (1986). (We apologize for the inversion of the currents on Figs. 2 and 4 of this paper.)

10. J.C. Manifacier, Y. Moreau and H.K. Henisch, "A tunable, Current-Controlled, Light-Emitting Diode," Solid State Electron 30:354 (1987).

11. J.C. Manifacier and H.K. Henisch, "Minority-Carrier Injection into Semiconductors Containing Traps," Phys. Rev. B 17:2648 (1978).

12. J.C. Manifacier and H.K. Henisch, "The Interpretation of Ohmic Behavior in Semi-Insulating GaAs Systems," J. Appl. Phys. 52:5195 (1981).

13. T. Stoica and C. Popescu, "Bulk Boundary Conditions for Injection and Extraction in Trap-Free Lifetime and Relaxation Semi-Conductors," Phys. Rev. B 17:3972 (1978).

14. M. Ilegems and H. Queisser, "Current Transport in Relaxation-Case GaAs," Phys.Rev. B 12:1443 (1975).

15. G.H. Dohler and H. Heyszenau, "Conduction in the Relaxation Regime," Phys.Rev. B 12:641 (1975).

16. J.C. Manifacier and H.K. Henisch, "The concept of Screening Length in Lifetime and Relaxation Semiconductors," J. Phys. Chem. Solids, 41:1285 (1980).

17. D. Scharfetter and H.K. Gummel, "Large Signal Analysis of a Silicon Read Diode," IEEE Trans. Electron Dev. ED-16:64 (1969).

18. W.L. Engl and H. Dirks, "Numerical Device Simulation Guided by Physical Approaches," Proc. N.A.S.E.C.O.D.E. International Conf., p.65, Boole Press, Dublin, (1979).

19. D.M. Caughey, "Simulation of U.H.F. Transistor Small Signal Behavior to 10 GHz for Circuit Modeling," Proc. 2nd Cornell Conf. Computerized Electron, p. 369 (1969).

20. Y. Moreau, "Contribution aux Modelisations des Contacts Metal-Semiconducteurs," These d'Etat, unpublished, Montpellier (France) (1983).

21. R.E. Bank, J. Rose and W. Fichtner, "Numerical Methods for Semiconductor Device Simulation," IEEE Trans. on Electron Dev., 9:1031 (1983).

22. M.A. Lampert and P. Mark, "Current Injection in Solids," Academic Press (1970).

... ...
... ...

A SEMICONDUCTOR MODEL FOR ELECTRONIC THRESHOLD SWITCHING

Peter T. Landsberg

Department of Electrical Engineering
University of Florida
Gainesville, FL 32611, USA[*]

In this short note I want to serve Heinz Henisch's interest in switching phenomena by giving a simple and nonmathematical description of one of my favorite models for electronic threshold switching.

Suppose one has one rate equation for the electron concentration in a semiconductor and one for the hole concentration, obtained, for example, from a study of a recombination-generation regime. Suppose, in addition, that the steady-state version of these two equations allows for four distinct steady-state solutions (n_i, p_i) ($i = 1, 2, 3, 4$). Two of the recombination or generation parameters may be regarded as under one's control. We have chosen the coefficient, X_1, governing the impact ionization of a deep trap by an electron and X_4, governing the impact ionisation of the same trap by a hole. It is possible to map the ranges of stability of the four steady-state solutions in the (X_1, X_4)-plane, which thus becomes a simple physical example of a control parameter space.

The control parameters X_1, X_4 depend strongly on the applied electric field. It follows that as the electric field is increased, the system describes a trajectory (AB say, Fig. 1) in the (X_1, X_4)-plane. This will in general pass through several of the regions of stability shown in this plane. As the trajectory AB crosses the boundary between two regions of distinct steady-state solutions (at a point P on Fig. 1), the system will switch. For example, if

$$n_1 = 0, \; p_1 = 0; \qquad n_2 = 0, \; p_2 = p' > 0 \; ,$$

then there will be a sudden increase in the hole concentration from 0 to p'. There will be no switch at Q, since the trajectory remains in region 2, but there will be another switch at R, since transition is made to solution 3.

Suppose the electric field is next lowered, so that the trajectory is traversed in precisely the opposite sense B → A. Now there is no transition at R, since the trajectory remains in region 3, but there are

---

[*]Permanent address: Department of Mathematics, University of Southampton, Southampton, England SO9 5NH.

switches at points Q and P as the stability regions 2 and 1 are entered, respectively. Since the points Q and R occur only at one of the two traversals, such a system exhibits hysteresis. This occurs because two regions of stability have been assumed to overlap in (X_1, X_4)-space. The theoretical situation is therefore remarkably simple, being based on steady-state solutions of equations of the type

$$\dot{n} = (a + bn + cp)n, \quad \dot{p} = (p + qx + rp)p$$

where a, b, c, p, q, r depend on standard recombination-generation coefficients.

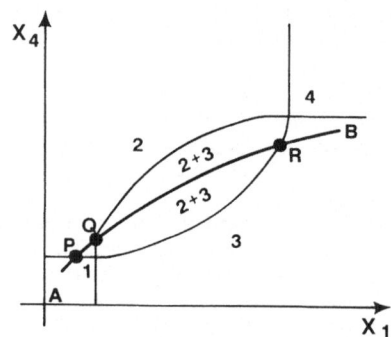

Fig. 1. The (X_1, X_4)-control parameter space showing four regions of stability for steady-state electron and hole concentrations.

Such a model can be made quantitative, using only standard numerical data from the semiconductor literature.[1] The results are in reasonable agreement with at least some experiments on amorphous silicon and amorphous selenium [ref. 2, Table 2]. Since every ingredient of this model - kinetic equations, recombination and generation coefficients, electric field dependences - are standard in semiconductor physics and susceptible to independent experimental study, it has a good chance of representing a broad class of switching transitions.

The historical background of this paper has been omitted since an introduction to this field and a survey of recent work with many appropriate references has just been published elsewhere.[3]

One can go a step further and ask if perhaps a simple approach of this type might also be available to discuss chaos in the form of irregular or aperiodic voltage or current oscillations in semiconductors, see ref. 3 and 4 and papers cited there. Interestingly enough, this is indeed possible,[5] but it is a different story.

REFERENCES

1. D.J. Robbins, P.T. Landsberg and E. Schöll, Phys. Stat. Sol. (a) <u>65</u>, 353 (1981)
2. P.T. Landsberg, D.J. Robbins and E. Schöll, Phys. Stat. Sol. (a) <u>50</u>, 423 (1978).
3. E. Schöll, <u>Nonequilibrium Phase Transitions in Semiconductors</u> (Berlin: Springer, 1987).
4. A. Brandl, T. Geisel and W. Prettl, Eurephys. Lett. <u>3</u>, 401 (1987).
5. P.T. Landsberg, E. Schöll and P. Shukla, Physica D (in the press).

PROPER CAPACITANCE MODELING FOR DEVICES

WITH DISTRIBUTED SPACE CHARGE

Roberto C. Callarotti

INTEVEP S.A.
Apartado 76343, Caracas 1070A
Venezuela

INTRODUCTION

The present paper derives from our previous work (1-3) on modeling the time dependence of ovonic threshold switches (OTS), and DIACS (4). It is also related to our analysis (5-8) of one carrier, space-charge controlled conduction in metal-insulator metal (MIM) structures.

The experimental work with OTS and DIACS was concerned with determining an equivalent circuit for the devices, derived from measurements of their relaxation oscillation behavior, or from their response to low-level sinusoidal excitation. In this type of work we did not attempt to derive theoretical expressions for the circuit elements of the equivalent circuits (except in the case of the resistances of the DIACS).

In the work with MIM structures we attempted to derive expressions for the capacitance of these devices, and we realized then that for many electronic devices with space charge, the models derived in the literature (9-11) were not formally correct.

We began then to work on the distributed capacitance equivalent circuit which is the subject of this paper, and which has been partially presented previously (12).

In Part I, we review the definition of capacitance from a circuital point view, as well as from electromagnetic theory. In part II we describe our distributed capacitance model for systems with space charge. In Part III we present the distributed model for a MIM and we compare our model with the numerical incremental impedance of the device. Finally in Part IV, we discuss possible implications of our distributed capacitance model for any device, and present our conclusions.

I)   WHAT IS CAPACITANCE?

This very simple question has a very simple answer from the circuital point of view, that is in terms of Kirchoff voltage and current laws. We _define_ a time invariant capacitance C, as a circuit element whose characteristic is given by:

$$I(t) = C \, d\bar{V}(t)/dt \qquad\qquad I\text{-}1$$

As clearly indicated in Figure 1, the charge Q_{21} which enters the device at terminal A in a time interval t_2-t_1, is the same as the charge which leaves the device at terminal B in the same time interval. This charge is related to the voltage difference $V(t_2)-V(t)$ by:

$$Q_{21} = \int_{t_1}^{t_2} I(t) \, dt \quad = \quad C\{\, V(t_2)-V(t_1)\} \qquad\qquad I\text{-}2$$

This result is obvious, as required by the Kirchoff current law (or conservation of charge). Furthemore if the charge which enters at A is positive, the charge which leaves at B is also positive, and with the same magnitude. Equivalently we can think of a negative charge entering the device at terminal B. Thus this "capacitor", in the given time interval has stored a positive charge at the upper electrode and a negative charge (of the same magnitude) at the lower electrode, with zero overall charge storage. What has really been stored is an electric field.

The behaviour of such a capacitor can be easily seen from the circuital equation I-1, if we apply an impulse of current:

$$I(t) = Q_0 \, u_0(t) \qquad\qquad I\text{-}3$$

where Q_0, the weight of the impulse, has magnitude of charge, and thus the charge and the voltage in the device are given by:

$$Q(t) = Q_0 u_{-1}(t) \qquad\qquad I\text{-}4$$

$$V(t) = (Q_0/C) \, u_{-1}(t) \qquad\qquad I\text{-}5$$

we have used Guillemin's notation (13) for the unit impulse and the unit step.

The impulse places a charge Q_0 at the upper electrode, and a charge $-Q_0$ at the lower one. If we define as V_0 the DC value of the voltage, then the capacitance C of the system has the usual meaning:

$$C = Q_0/V_0 \qquad\qquad I\text{-}6$$

This relation for capacitance in terms of charge can only be valid when two charges of equal magnitude and different sign exist at the two electrodes of the device. As we will shortly see from the electromagnetic analysis of a structure of two parallel metal plates, the charges $+Q_0$ and $-Q_0$ will be located at the inner faces of the metal places (in the case of a metal with infinite conductivity), in the form of surface charges of magnitude $\eta = Q_0/A$, where A is the surface of the plates.

Figure 2 illustrates the structure considered. The proper electromagnetic analysis of the system requires the solution of the full set of Maxwell's equations:

$$\vec{\nabla} \times \vec{E} = -\partial \vec{B}/\partial t \qquad\qquad I\text{-}7$$

$$\vec{\nabla} \times \vec{H} = \vec{J} + \partial \vec{D}/\partial t \qquad\qquad I\text{-}8$$

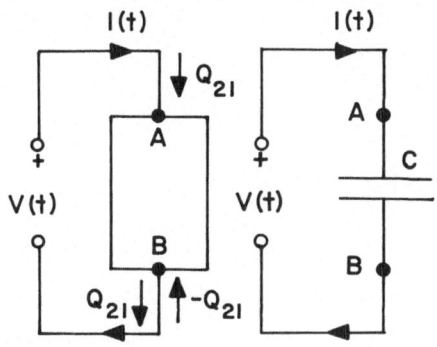

Fig. 1. Circuital definition
of capacitance

$$\vec{\nabla} \cdot \vec{D} = \rho \qquad\qquad \text{I-9}$$

$$\vec{\nabla} \cdot \vec{B} = 0 \qquad\qquad \text{I-10}$$

and the corresponding constituitive relations:

$$\vec{J} = \sigma \vec{E} + \vec{J}_{diff} \qquad\qquad \text{I-11}$$

$$\vec{D} = \varepsilon \vec{E} \qquad\qquad \text{I-12}$$

$$\vec{B} = \mu \vec{H} \qquad\qquad \text{I-13}$$

where we have introduced the conductivity σ, the permittivity ε, and the permeability μ, of the medium between the plates. J_{diff} is the diffusion current which is present when gradients of carriers densities exists.

In the above equations ρ represents the volume density of free charges.

We will not solve these equations in detail here, but we refer the reader to references 14 and 15, where they are solved in the quasistatic limit, which is particularly pertinent for the purposes of this paper.

For the structure considered in Figure 2, the quasistatic zeroth order solution yields ($\sigma = 0$, $\rho = 0$):

$$\vec{H}_o = 0 \qquad\qquad \text{I-14}$$

$$\vec{E}_o = -V_o(t) /d \; \vec{i}_x \qquad\qquad \text{I-15}$$

$$\vec{J}_o = 0 \qquad\qquad \text{I-16}$$

Application of the boundary conditions for E at the upper and lower electrodes (the discontinuity of the electrical density D) yields the surface charges at the electrodes:

$$\eta (d) = V(t)/d \qquad\qquad \text{I-17}$$

$$\eta (0) = - V(t)/d \qquad\qquad \text{I-18}$$

so that the total charge at the plates of area A, are:

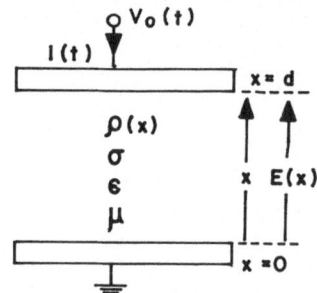

Fig. 2. Two metal plates
limiting a given
medium.

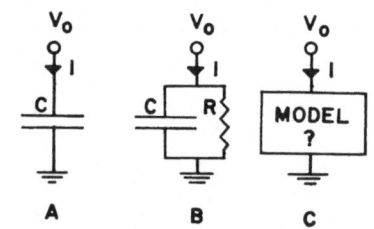

Fig.3. Circuit models for
the structure of Fig. 2
(a)$\rho = 0$ $\sigma = 0$
(b)$\rho = 0$ $\sigma \neq 0$
(c)$\rho \neq 0$ $\sigma \neq 0$

$$Q(d) = + \quad A \, V(t)/d \qquad\qquad I-19$$

$$Q(0) = - \quad A \, V(t)/d \qquad\qquad I-20$$

As we can see we have a device with two charges of equal magnitude and opposite sign, "separated" by a voltage $V(t)$, so the we can define a circuital capacitance given (with no great surprise) by:

$$C = Q(d)/V(t) = \varepsilon A/d \qquad\qquad I-21$$

The equivalent circuit is shown in Fig. 3a.

If we consider the situation depicted in Figure 2, where the plates are separated by a medium with conductivity $\sigma \neq 0$, the zeroth order analysis yields the following answers:

$$\vec{H}_o \neq 0 \qquad\qquad I-22$$

$$\vec{E}_o = - V_o(t)/d \; \vec{1}_x \qquad\qquad I-23$$

$$\vec{J}_o = \sigma \, V_o(t)/d \; \vec{1}_x \qquad\qquad I-24$$

The surface charges will be the same, since we have the same electric field, so that the capacitance can be defined properly, and it will have the same value as before (Eq. I-21). Additionally we have a current flowing under the voltage excitation, given by:

$$I(t) = A \left| J_o \right| = \sigma \, A \, V_o(t) \, /d \qquad\qquad I-25$$

The above relation defines a resistance R given by:

$$R = V_o(t)/I(t) = d/(\sigma A) \qquad\qquad I-26$$

Equation I-8 gives us the key as to how to combine the two circuit elements derived, since it shows that there are two currents flowing due to the applied voltage, one conductive (J_o) and one displacement (dD/dt) associated to the derived capacitance. Obviously the two circuit components should be connected in parallel, so that the equivalent circuit obtained is that shown on Figure 3b.

Before we go on let us stress that the obtained value of resistance and capacitance are the same than those that would have been obtained under applied DC voltages.

This is the essence of the quasistatic method which allows us to derive equivalent circuits which are valid for a certain range of time variations, in terms of results derived in the limit:

$$\partial / \partial t \rightarrow 0 \qquad (or \quad w \rightarrow 0) \qquad\qquad I\text{-}27$$

In order to determine the range of frequencies w, for which the equivalent circuits to be presented in this paper are valid, we must find the solution of higher order quasistatic approximations; this we will leave for future discussions.

The problem becomes more complex when we consider the situation illustrated in Figure 6, where the medium between the plates has free charges present ($\rho \neq 0$). We will derive the corresponding model (Figure 3c).

Lets consider the DC solution of a pair of metal plates separated by a distance d, enclosing a space charge distribution $\rho (x)$. Inside the plates we must solve:

$$\vec{\nabla} . \vec{E} = \rho / \varepsilon \qquad\qquad I\text{-}28$$

Under DC conditions, we can define a potential V(x) satisfying:

$$\vec{E} = - \vec{\nabla} V \qquad\qquad I\text{-}29$$

so that inside the plates we must solve Poisson's equation:

$$\nabla^2 V(x) = - \rho / \varepsilon \qquad\qquad I\text{-}30$$

by integrating twice we obtain:

$$V(x) = - \int_0^x \int_0^x \rho(x) \ d^2 x \ / \varepsilon \ + C_1 x + C_2 \qquad\qquad I\text{-}31$$

where C_1 and C_2 are integrating constants which are evaluated to be:

$$C_2 = 0 \text{ from } V \ 0)=0 \qquad\qquad I\text{-}32$$

$$C_1 = V_o/d + (1/d) \int_0^d \int_0^x \rho(x) \ dx^2 \ / \varepsilon \qquad\qquad I\text{-}33$$

from the boundary condition V(d) $=V_o$. Thus the potential and the electric field are given by:

$$V(x) = x \ V_o/d - \int_0^x \int_0^x \rho(x) \ dx^2 \ / \varepsilon \ + (x/d) \int_0^d \int_0^x \rho(x) \, dx^2 / \varepsilon \quad I\text{-}34$$

$$E(x) = -V_o/d + \int_0^x \int_0^x \rho(x) \, dx / \varepsilon \ - (1/d) \int_0^d \int_0^x \rho(x) \, dx^2 / \varepsilon \qquad I\text{-}35$$

105

On the bases of the values obtained for the electric field we can evaluate the charges at the upper and lower plates:

$$Q(0) = - AV_0/d - (A/d) \int_0^d \int_0^x \rho(x)\,dx^2 \qquad \text{I-36}$$

$$Q(d) = + AV_0/d + (A/d) \underbrace{\int_0^d \int_0^x \rho(x)\,dx^2 - A \int_0^d \rho(x)\,dx}_{} \qquad \text{I-37}$$

$\underbrace{}$

due to the external   due to the internal
potential   space charge

This structure <u>cannot</u> be represented as a capacitor since:

$$Q(0) \neq - Q(d) \qquad \text{I-38}$$

This result should not be surprising since the system that we are modeling has distributed charges, and it would be indeed strange if it could be modeled as a lumped single resistor-condenser circuit as in the case of Figure 3b. What we should expect is a distributed parameter model with an infinite number of circuit elements, and indeed this is the case that we will encounter in the next section, where we derive the proper model for the case discussed above.

II) DISTRIBUTED CAPACITANCE MODEL

If we consider a special case of the example previously discussed, where the space charge is constant $\rho(x) = \rho_0$, we have:

$$Q(0) = - \varepsilon AV_0/d - A \rho_0 d/2 \qquad \text{II-1}$$

$$Q(d) = + \varepsilon AV_0/d - A \rho_0 d/2 \qquad \text{II-2}$$

In this case if we could say that:

$$\varepsilon AV_0/d \gg A \rho_0 d/2 \qquad \text{II-3}$$

then:

$$Q(0) \cong - Q(d) \qquad \text{II-4}$$

and we will be approaching a proper capacitor with almost equal and opposite charges on each plate.

We can rewrite II-3 as follows:

$$2 \varepsilon V_0 / \rho_0 \gg (d)^2 \qquad \text{II-5}$$

Equation II-5 is the key for the solution of the modeling of a system with $\rho(x)$, and $E(x)$, such as that shown on Figure 4.

We begin by slicing the devices in slices of equal thickness dx, so that the voltage across each of them is dV. If the slices are thin enough II-5 is satisfied:

$$2 \varepsilon \, dV / \rho(x_o) \gg (dx)^2 \qquad\qquad \text{II-6}$$

Obviously we can make the slices thin enough so that II-6 can always be satisfied. Once we do this, from the zeroth order quasistatic solution of Maxwell equation we determine the slice capacitance which is given by:

$$C_i = \varepsilon \, A/dx \qquad\qquad \text{II-7}$$

Now we have to find other circuit components that complete the model for the slice. This is easy since we know that we can look at the DC solution $E(x)$ field in the slice. As we will show in the next section, in the application of the model to a MIM structure, we are interested in incremental linear models, since the basic equations are non-linear. Thus in the case of incremental $E(x)$, DC fields, we can write:

$$E(x) = f(x) \, J = f(x) \, I/A \qquad\qquad \text{II-8}$$

Where I is the incremental applied current (which is constant through the device). J is the current density and we have defined $f(x)$ as the incremental electric field per unit applied incremental current density. If the slice is thin enough to satisfy II-6, and thin enough that the electric field across it can be considered equal to the constant value $E(x_i)$, then:

$$V_i/dx = E(x_i) \qquad\qquad \text{II-9}$$

Where the sign is due to the convention used in the definition of the x axis in Figure 4. Using II-8 in II-9 yields:

$$V_i = f(x_i) \, dx \, I \, /A \qquad\qquad \text{II-10}$$

and we realize that II-10 is the equation for a resistance R_i, which characterizes the slice conduction, given by:

$$R_i = f(x_i) \, dx \, /A \qquad\qquad \text{II-11}$$

and from our definition in II-8 we obtain:

$$f(x_i) = E(x_i)/J \qquad\qquad \text{II-12}$$

V_i in the above equations is the incremental voltage difference across each slice.

The procedure described above is illustrated in Figure 4 and in Figure 5.

It is proper at this point to state, that the assumption of constant E field across the slice, as well as the assumption of constant space charge do not imply that we are disregarding diffusion, or any other current. The values of incremental $E(x)$, and $n(x)$, that we use in the derivation of the circuital models, are the values obtained from the numerical solution of the complete equations which describe the system, which include diffusion terms. This fact will become clearer in the next section, where we will derive the equivalent circuit for a MIM structure.

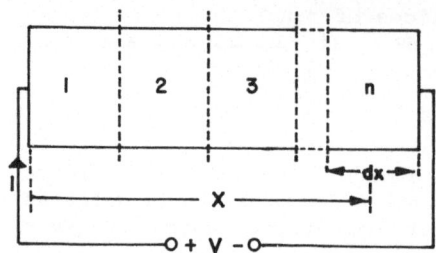

Fig. 4. Separation of a device
with space charge.

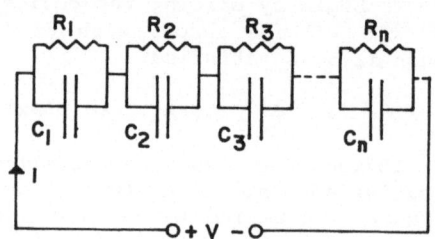

Fig. 5. Corresponding
distributed capacitance
equivalent circuit

Before we go on, it is convenient to consider the total impedance of
the device that we have modeled. This total impedance is the sum of the
impedance of each slice. In terms of the Laplace transform frequency s,
we have, that for the case of n equally spaced slices, the total incremental impedance Z(s), is given by:

$$Z(s) = \sum_{i}^{n} R_i / (1 + sR_i C_i) \qquad \text{II-13}$$

where all the C_i's are equal and given by II-7.

In the special case where all the slices have equal R_i's, the
incremental electric field is constant in space, so that:

$$Z(s) = nR_i / (1+sR_i C_i) \qquad \text{II-14}$$

we can rewrite this result as:

$$Z(s) = nR_i / (1+ (nR_i)(C_i/n)s) \qquad \text{II-15}$$

so that the system is equivalent to a one RC system (a one time constant
system), with: II-16

$$R_{lumped} = nR_i = \text{the total DC resistance}$$
$$\text{of the system} \qquad \text{II-17}$$

$$C_{lumped} = C_i/n = \text{the total C of the system}$$

if we use II-7 in II-17 we obtain: II-18

$$C_{lumped} = \varepsilon \ A/(n \ dx) = \varepsilon A/L \equiv C_u$$

where L is the length of the device, showing that the lumped capacitance
is the capacitance of the whole device   We define this "uniform"
capacitance as C_u, corresponding to the "geometrical" capacitance of
the device.

In the general case where the R_i's are different, one can easily
show that in the limit of small s, Z(s) tends to:

$$Z(s) \underset{s \to 0}{\to} \to \cdot \quad \frac{\sum\limits_{i}^{n} R_i}{1 + \{\sum\limits_{i}^{n} R_i\} C_o \, s} \qquad \text{II-19}$$

where C_o is given by:

$$C_o = C_i \, \sum\limits_{i}^{n} R_i^{2} \Big/ \{\sum\limits_{i}^{n} R_i\}^{2} \qquad \text{II-20}$$

C_o is the capacitance that would be measured by incremental AC techniques in the low frequency region ($s = jw$).

In Appendix B, we will determine the value of C_o for different field distributions.

III) APPLICATION OF THE MODEL TO A MIM

In order to determine the comparison of the distributed capacitance model to other models previously used for the calculation of the capacitance of a device, we will consider a MIM structure.

We will use the normalization used in previous work (5-6), and we will consider a system without traps. The capacitance will be normalized according to:

$$C_{normalized} = C_{real} \, / \, \varepsilon \qquad \text{III-1}$$

We will consider system with unit cross sectional areas (A=1), so that according to II-7 the normalized C_i will be:

$$C_i = 1/dx \qquad \text{III-2}$$

Figure 6 shows E(x), and N(x) for J=10, for a one-dimensional MIM structure. Figure 7 shows the incremental electric field calculated numerically and analytically (as discussed in Appendix A).

The value of the normalized capacitance of each slice C_i, for a sample with n slices and normalized unit length, is given by:

$$C_i = n \qquad \text{III-3}$$

The equivalent circuit is thus completely determined, and the total impedance Z(s) can be calculated, as shown on Figure 8, as a function of the Laplace parameter s. The total impedance as a function of an AC frecuency w, can be simply evaluated by using s=jw in the impedance of each slice:

$$Z(w) = \sum\limits_{i}^{n} R_i \, / \, (1 + jw \, R_i \, C_i) \qquad \text{III-4}$$

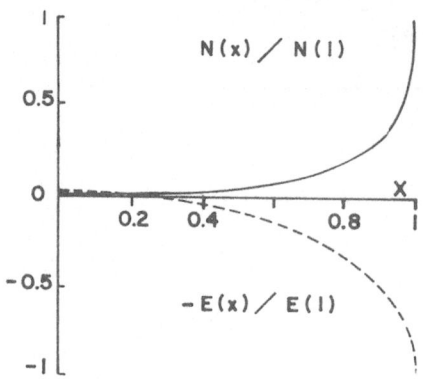

Fig. 6. Normalized DC charge
density N(x) and
electric field E(x) for a
MIM structure with:
N(O) = 1, N(1) =100, J = 10

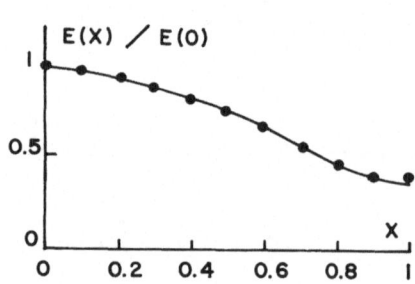

Fig. 7. Normalized DC incremental electric field for
the MIM structure of
Fig. 6: continous line
numerical solution;
dots analytical
solution.

The resistance Ri for each slice is obtained from the exact incremental electric field in each slice, according to equation II-11 and II-12.

IV)   COMPARISON WITH THE MIM IMPEDANCE

In order to check the results of our distributed capacitance model, we calculated directly the total incremental Z(s) impedance of the MIM. We achieved this by solving the linearized MIM equations numeracally.

$$j = E_O(x) \; n(x) + N_O(x) \; e(x) + dn(x)/dx + s \; e \; (x) \qquad \text{IV-1}$$

$$de(x)/dx = -n(x) \qquad \qquad \text{IV-2}$$

Where the lower case letters indicate incremental values, and the upper case letters indicate the large signal solutions. We solved the incremental equations numerically by the shooting method, with the following boundary conditions.

$$n(O) = -de(x)/dx \Big|_{x=0} = 0 \qquad \qquad \text{IV-3}$$

$$n(1) = -de(x)/dx \Big|_{x=1} = 0 \qquad \qquad \text{IV-4}$$

We found the values of n(x), e(x), and from the electrical field we computed the incremental voltage difference across the device, (which we define as v_{10} (s)) when the system is excited by a current density j(s). The incremental total impedance Z(s) is defined as:

$$Z(s) = v_{10}(s)/j(s) \qquad \qquad \text{IV-5}$$

for a system with unit cross sectional area.

Figure 8 shows the results of IV-5, where s was considered real, plotted as solid points while he continuous curve is given by the RC distributed circuit described in the previous section. The figure clearly shows the excellent agreement obtained.

V) CONCLUSIONS

As we have shown, the derived distributed capacitance equivalent circuit in particularly simple. We summarize here the procedure followed in the derivation:

a) Divide the sample with space charge, in thin enough slices so that the charges induced by the voltage across the slice are much larger than other charges (so that II-6 is satisfied).

b) The equivalent circuit for each slice is very simple, and it consist of a parallel combination of a capacitance C_i, and a resistance R_i.

c) If the device has uniform permittivity, and if the slices all have the same width, then each of the slice capacitances are equal, with a value given by II-7.

d) The values for the different R_i's, are found from the incremental DC solution for the electric field in the device according to II-11 and II-12.

e) The total impedance for the device can then be found from the equivalent circuit, by summing the impedance of each slice.

f) In the case of incremental AC measurements, where we apply a sinousoidal incremental current at a frequency w, in the limit of zero frequency, the systems will behave as a one time constant system with and equivalent total resistance given by the sum of the slice resistances, and an equivalent lumped capacitance C_0 given by II-20.

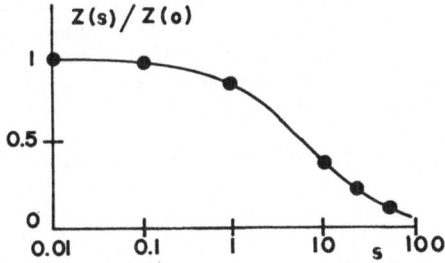

Fig. 8. Normalized impedance for the MIM of figures 6 and 7 vs. s (considered real). Continous curve gives results of the distributed capacitance model. Dots indicate numerical solutions of the equations IV-1 and IV-1.

In Appendix B, we compare the values of capacitance in the low frequency limit, as obtained from the equivalent circuit described above, with the equivalent capacitance defined in terms of the electric field at one electrode (as we used in our work in reference 5). We carry out the comparison for several cases of assumed electric field distributions.

Finally we would like to state that the equivalent circuit procedure presented in this paper should be very helpful in determining proper capacity values for a variety of devices, from solar cells, to MOS structure, ballistic devices, Schottky barriers, diodes, etc.

Acknowledgement

The author would like to express his gratitude to Dr. Heinz K. Heinsch, for a very fruitful collaboration of many years. Third World science might even have a chance if collaborative efforts similar to those that have been induced by Henisch, could occur in numerically significant instances.

APPENDIX A ANALYTICAL SOLUTION FOR THE INCREMENTAL MIM (DC)

In the usual normalized from (5,6) the MIM equations are:

$$J = N(x)E(x) + dN(x)/dx \qquad\qquad \text{A-1}$$

$$dE(x)/dx = - N(x) \qquad\qquad \text{A-2}$$

If we assume that:

$$J = J_O + j \quad \text{,with} \quad J_O \text{ much large that } j \qquad\qquad \text{A-3}$$

Then we can write:

$$N(x) = N_O(x) + n(x) \qquad\qquad \text{A-4}$$

$$E(x) = E_O(x) + e(x) \qquad\qquad \text{A-5}$$

$$V(x) = V_O(x) + v(x) \qquad\qquad \text{A-6}$$

Where the lower case letters indicate incremental values, V, V_O and v represent voltages($E(x) = -dV(x)/dx$). The incremental values are assumed to be small in comparison with the upper case zero suscripted values, which represent the bias point value of the variables. Substituting A-3 to A-6 in A-1 and A-2, and keeping first order terms only, yields the incremental equations:

$$J = N_O(x)\ e(x) + E_O(x)\ n(x) + dn(x)\ /dx \qquad\qquad \text{A-7}$$

$$de(x)/dx = - n(x) \qquad\qquad \text{A-8}$$

They represent a system of linear equations with varying coefficients, which we will now proceed to solve analytically.

First we obtain an equation for $e(x)$, by substituting A-8 in A-7:

$$j = N_o(x) \ e(x) - E_o(x) de(x)/dx - d^2 e(x)/dx^2 \qquad \text{A-9}$$

and using the zero suscripted version of A-2:

$$j = - \{ e(x) \ dE_o(x)/dx + E_o(x) \ de(x)/dx + d^2 e(x)/dx^2 \} \quad \text{A-10}$$

which we can rewrite as:

$$j = - \ d/dx \ \{ E_o(x) \ e(x) + de(x)/dx \} \qquad \text{A-11}$$

and by integration:

$$-(jx + b) = E_o(x) \ e(x) + de(x)/dx \qquad \text{A-12}$$

where b is an integration constant to be evaluated from boundary conditions. Equation A-12 has the following solution:

$$e(x) = \exp (V_o(x) \ \{ a - \int_o^x \{\exp(-V_o(y))\} \ (jy + b) \ dy\} \quad \text{A-13}$$

where a is a new integration constant. Using A-8 and A-13 we can show that:

$$n(x) = jx + b + E_o(x) \ e(x) \qquad \text{A-14}$$

by integrating A-13 by parts, and introducing the values A1(x) and A2(x):

$$A1(x) = \int_o^x \exp(-V_o(y)) \ dy \qquad \text{A-15}$$

$$A2(x) = \int_o^x \int_o^x \exp(-V_o(y)) \ dy \ dy \qquad \text{A-16}$$

we can rewrite A-13 as follows:

$$e(x) = \exp(V_o(x) \ \{ a - (jx + b) \ A1(x) + j \ A2(x) \} \quad \text{A-17}$$

the boundary conditions are:

$$n(0) = 0 \quad \rightarrow \quad - E_o(0) \ e(0) = b \qquad \text{A-18}$$

$$n(1) = 0 \quad \rightarrow \quad - E_o(1) \ e(1) = b + j \ (1) \qquad \text{A-19}$$

and the constants a and b can be evaluated as:

$$a = j \ \frac{\exp(-V_o(1) - \ \{A1(1) - A2(1)\} \ E_o(1)}{E_o(0) \ \exp(-V_o(1) - E_o(1)\{ \ A1(1) \ E_o(0) + 1 \ \}} \qquad \text{A-20}$$

$$b = - \ a \ E_e(0) \qquad \text{A-21}$$

As mentioned before, Figure 7, shows the comparison between these analytic solutions and the direct numerical calculation for the incremental MIM (DC).

COMPARISON OF THE LOW FREQUENCY DISTRIBUTED CAPACITANCE WITH THE CAPACITANCE EVALUATED FROM THE FIELD AT $x=0$ FOR DIFFERENT SYSTEMS.

In order to evaluate the differences between the capacitance predicted by the ditributed capacitance model (in the w → 0 limit) and the capacitance determined from the field at $x=0$, we consider with resistance with the following x dependence.

$$R(x) = A_0 \exp(-x/alfa) \qquad \qquad \text{B-1}$$

Recalling II-11 and II-12, this will be the x dependence of the corresponding e(x) field. We adjust A_0 so that the value of the impedance of the system at zero frequency (w=0) will always be the same, and we will look at cases corresponding to different values of alfa. If alfa is much larger than 1, the field e(x) will be essentially constant through the sample. On the other hand, if alfa is much smaller than 1, the field will vary strongly with x, over the length of the sample. This behavior is illustrated on Figure 9.

Lets define Zrc as the impedance given by the distributed capacitance model in the w → 0 limit. Lets define Co as the corresponding capacitance.

Lets define Zp as the impedance obtained by evaluating the pseudo capacitance of the system in the manner indicated in reference 8, from the value of e(0). We define Cp as the corresponding capacitance.

We define Cu as the capacitance (normalized) corresponding to a device with n equal slices, each with capacitance C_i given by III-2. Thus the normalized Cu value will be 1, for a sample of unit normalized length.

We have:

$$\int_0^1 R(x)\, dx = Dx \sum_i^n R_i(x) \qquad \qquad \text{B-2}$$

from the definition of an integral. Using B-1, and III-4, we have:

$$A_0 = Dx\, Zrc(0) \,/\, \{\, alfa\, (1 - \exp(-2/alfa))\,\} \qquad \text{B-3}$$

where Dx is the finite width of each slice, and for proper comparison we take Zrc(0)=Zp(0).

The value of CP is found in the following way:

$$Cp = e(0)/v(1) \qquad \qquad \text{B-4}$$

where the DC voltage across the device is given by:

$$v(1) = j\, Zp(0) = j \sum_i^n R_i \qquad \qquad \text{B-5}$$

and the electric DC incremental a field at $x=0$ is given by:

$$e(0) = R(0)\, j\, /\, Dx = j\, A_0\, /\, Dx \qquad \qquad \text{B-6}$$

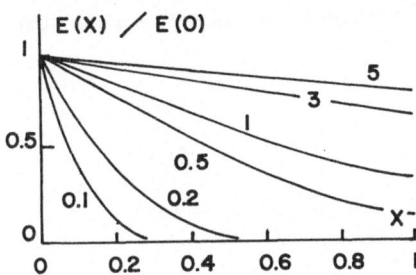

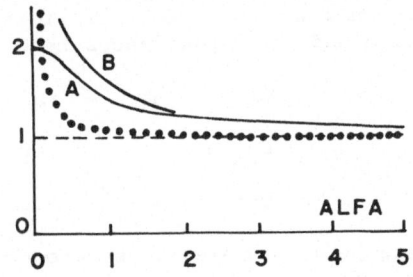

Fig. 9. Assumed incremental DC
field dependences vs.
normalized distance.
Values on the curves
refer to values of the
parameter alfa.

Fig. 10. Comparison of different
capacitances in the low frequency
limit: (A) Cp/Co (B) Cp/Cu and
dots Co/Cu, as functions of the
parameter alfa.

where we have used B-1. Thus we find:

$$Cp = A_o / \{ Dx \sum_i^n R_i \} \qquad \text{B-7}$$

and from B-3 we obtain:

$$Cp = Cu / \{ \text{alfa} (1 - \exp(-1/\text{alfa})) \} \qquad \text{B-8}$$

Evaluating II-20, in terms of the different dependences given by
B-1, will allow us to calculate the value of Co, the capacitance given by
the distributed RC model in the w → 0 limit:

$$Co = Cu \, Dx \sum_i^n R_i^2 / \left[Dx \sum_i^n R_i \right]^2 \qquad \text{B-9}$$

and evaluating the sums in terms of R(x) and $R^2(x)$ integrals:

$$Co = Cu \, \frac{\{ 1 - \exp(-2/\text{alfa}) \}}{2 \, \text{alfa} \, \{ 1 - \exp(-1/\text{alfa}) \}^2} \qquad \text{B-10}$$

and by substituting the appropiate limits for alfa, we can determine
the following relations:

$$\text{alfa} \to \infty \qquad Cp \to Co \to Cu \qquad \text{B-11}$$

where this case corresponds to a constant or linearly varying e(x)
incremental DC field across the device. In this case the distributed
capacitance at zero frequency tends to the pseudo capacitance value, and
both tend to the capacitance Cu, the "geometrical" capacitance of the
device.

In the alfa $\rightarrow$ O limit, the field varies strongly across the device and the capacitances tend to:

$$Cp \rightarrow 2\ Co \qquad\qquad B-12$$

so that:

$$Co \rightarrow Cu/(2\ alfa) \qquad\qquad B-13$$

$$Cp \rightarrow Cu/alfa \qquad\qquad B-14$$

these behaviors are illustrated on Figure 10.

REFERENCES

1) R. C. Callarotti and P. E. Schmidt, Amorphous and Liquid Semiconductors, W.E. Spear Ed. (Center for Industrial Consultancy and Liason), The University of Edinburgh, Edinburgh, 717-721 (1977).

2) P. E. Schmidt and R.C. Callarotti, "Theoretical and Experimental Study of the Operation of Ovonic Switches in the Relaxation Oscillation Mode: I- The charging characteristics", J. Appl. Phys., 55(8), 3144-3147, (1984).

3) R.C. Callarotti and P.E. Schmidt, "Theoretical and Experimental Study of the Operation of Ovonic Switches in the Relaxation Oscillation Mode: II- The discharge characteristics" J. Appl. Phys., 55(8), 3148-3152 (1984).

4) R.C. Callarotti et al., Circuit Theory for the Comparison of Relaxation Oscillations in OTS and DIACS, J. of Non Cryst. Solids., 35 & 36, 1117-1122 (1980).

5) H.K. Heinsch et al., "Space Charge Controlled Conduction in Thick MIM Barriers", J. Appl. Phys., 51, 3790-3793 (1980).

6) P.E. Schmidt et al., "Single-Carrier Space Charge Controlled Conduction", J. Appl. Phys., 53(7), 4996-5005 (1982).

7) P.E. Schmidt et al., "On the Resistance of Thin Insulating Films Exhibiting Contact-Limited Currents", Thin Solid Films, 89, 83-88 (1982).

8) U. K. Reddy et al., "Space Charge Controlled Pseudo-Capacitance in Thin Films", Solid State Electr., 28(5), 532-535, (2984).

9) M.A. Lampert and P. Mark, Current Injection in Solids, Chapter III, Academic Press, NY, (1970).

10) J.G. Simmons and G.W. Taylor, "Generalized Theory of Conduction in Shottky Barriers", Solid State Electr., 16(7), 705-709 (1983).

11) G. Gildenblat and S.S. Cohen, "On the nonequilibrium Capacitance of the Shottky Diode", J. Appl. Phys., 58(1), 607-608, (1985).

12) R.C. Callarotti, "Circuito Equivalente para Condensadores con Carga Espacial Interna y Cargas Superficiales Desiguales" (Equivalent Circuit for Capacitors with Internal Space Charges and Unequal Surface Charges), 32nd Annual Convention of the Venezuelan Society for the Advancement of Science (ASOVAC), Caracas (1982).

13) E.A. Guillermin, <u>Theory of Linear Physical Systems</u>, Chapter X, Wiley, NY, (1963).

14) R.M. Fano, L.J. Chu, and R.B. Adler, <u>Electromagnetic Fields Energy and Forces</u>, Chapter VI, Wiley, NY, (1960).

15) L.M. Magid, <u>Electromagnetic Fields, Energy and Waves</u>, Chapter IX, Wiley, NY,(1972).

[ ] . . . .

ON THE IMPEDANCE CALCULATION OF THICK MIM BARRIERS

P.E. Schmidt and R.C. Callarotti*

Instituto Venezolano de Investigaciones Científicas
Apartado 21827, Caracas 1020A, Venezuela
*Intevep S.A. Apartado 76343. Caracas 1070A. Venezuela

Methods for the calculation of the incremental a.c. characteristics of the thick metal-insulator-metal (M_1-I-M_2) barrier are described. A general approach resulting in the formulation of a set of two coupled ordinary differential equations in terms of the real and imaginary parts of the complex a.c. carrier density function is explained in detail. As an alternative and more practical method, the calculation of the driving-point impedance function of the M_1-I-M_2 barrier leading directly to an incremental equivalent circuit representation is given. It is found that the device may be modeled satisfactorily in the low frequency range for all bias conditions by a resistor connected in parallel with a capacitor.

INTRODUCTION

The d.c. (direct current) single-carrier space-charge controlled conduction in insulators with and without traps has been studied in great detail by many researchers[1]. Mott and Gurney[2] proposed a simplified theory for an insulator without traps resulting in the well-known square law

$$j = \frac{9}{8} \varepsilon \mu \frac{v^2}{1^3}$$ (1)

Where j is the current density, v is the applied voltage, ε is the dielectric constant of the insulator, μ is the electron mobility, and 1 is the insulator thickness. To obtain above expression the authors neglected the carrier difussion current. Further, one has to assume that the electric field at the plane of the injecting contact is equal to zero, which requires an infinite carrier concentration at that boundary in order to impose current continuity everywhere. Finally, the properties of the second contact of the device under study is considered to be of no importance to the solution of the problem. In orther words, the length of the device is very large with respect to the sum of the widths of the metal-insulator space-charge regions located in the insulator. As a consequence, the square law is dictated solely by the bulk properties of the device. The exact theory that takes into account diffusion has been solved analytically by Fan[3] who obtained a solution in terms of Airy integrals from which the d.c. I-V characteristics can be calculated over a limited range of currents. Shockley and Prim[4], and Wright[5] investigated single-carrier space-charge controlled conduction in n[+] i n[+] semiconductor devices, while Skinner[6] studied the case of the insulator. All these authors obtained analytical solutions in terms of modified Bessel functions of fractional order, while Lindmayer[7] obtained a series solution. Numerical solutions have been given by Lampert and Edelman[8] for the problem of a semi-infinite insulator with traps. The computed solution is valid over a limited regios (LR) in contact with the injecting contact and is matched to the approximate pure drift analytical solution from the end of LR to infinity. A novel methods for obtaining numerical solutions to the single-carrier controlled conduction problem including diffusion has been described by Bonham and Jarvis[9] for the case of infinite charge densities at both metal-insulator interfaces of the device. More recently, Henisch et al[10] investigated single-carrier space-charge controlled conduction in what they called thick metal-insulator-metal (M_1-I-M_2) barriers meaning that the thickness of the insulator is such that the space-charge regions located in the insulator at each metal-insulator interface touch or even may overlap each other. A distinct barrier is then formed even without traps. The authors obtained numerical solution of the drift-diffusion equation yielding electron energy and concentration contours for such thick M_1-I-M_2 barriers and their appropiate d.c. current voltage relationships over an extended range of currents (9 decades) for different device parameters such as insulator thickness, trap densities, etc.[11]

In contrast to the abundance of publications on the d.c. characteristics of single-carrier space-charge controlled conduction, the investigation of the incremental a.c. characteristics of the same device structures has received much less attention and only a few reports have been published. Shao and Wright[12] calculated the incremental admittance of the space-charge-limited dielectric diode and showed that for all frequencies this device may be represented by a conductance and susceptance in parallel. Also Yoshimura[13] investigated the incremental a.c. characteristics of single - carrier-controlled conduction in semiconductor punched-through p-n-p diodes. In both papers the authors neglected the diffusion term in the current density expression which is then described solely by the sum of the drift and displacement currents. By making this approximation, the current can be written now as the total time derivative of the electric field and is easily integrated. The resulting analytical solution is expressed in terms of the carrier transit time. The small-signal theory of transit time effects for electrons in solids, which results from this treatment, is analogous to the one for electrons in vacuum tubes that has been worked out by Llewellyn[14,15] in his classic papers on vacuum tube electronics at ultra-high frequencies. Alternatively, the standard complex number method of alternating current theory was used by Wright[16] to obtain an analytical description of the small-signal characteristics of the punched-through diode operating under low-field conditions and neglecting diffusion.

In thick M_1-I-M_2 barriers, the distribution of the carriers in the insulator is highly non-uniform and changes with the d.c. applied current such that the assumption of neglecting the diffusion term in the current expression is totally inappropiate. To calculate correctly the incremental admittance of such a device different approaches may be pursued. Callarotti[17], in the preceding paper, develops a distributed capacitor method resulting in an equivalent circuit composed of a series connection of cells consisting of a resistor in parallel with a capacitor. Each cell represents a slice of the dielectric material having a capacitance and an appropriately calculated paralled resistance. The method discusses a procedure to divide the insulator in a sufficient number of slices with identical width all having the same capacitance value, namely, the geometric capacitance of the slice. The circuital model this obatined is analogous to the driving-point impedance representation of a one port, linear, passive, reciprocal, lumped and time-invariant network using RC Foster components[18].

In this paper we will discuss two different methods. The first method is a general approach which leads to a system of two coupled ordinary differential equations of the second order from which the real and imaginary parts of the device admittance may be calculated. In the second method the driving-point impedance is calculated and plotted versus real values of the Laplace variable s. From the the plotted results, one can, in principle, synthesize the one-port network representing the equivalent circuit model of the device for a given bias point. This last method has been applied to the thick M_1-I-M_2 barrier.

THEORY

1. General Approach

The equations describing single-carrier space-charge controlled conduction in insulator films without traps are

$$j = q\mu nf + qD \frac{\partial n}{\partial x} + \varepsilon \frac{\partial f}{\partial t} \tag{1}$$

and

$$\frac{\partial f}{\partial x} = - \frac{qn}{\varepsilon} , \tag{2}$$

with

$$f = - \frac{\partial v}{\partial x} . \tag{3}$$

The first equation is the total current density which is the sum of the total conduction current density and the displacement current density. The total conduction current consists of the drift current and the diffusion current. Low field condition as been assumed such that the motion of electrons mav be described bv means of a constant diffusion coefficient D and a corresponding constant mobility μ. Equation (2) is Gauss' Law in differential form where f is the electric field. It is assumed that the insulator has no free resident carriers of its own. Further, the carriers considered in this study are arbitrarily taken to be electrons. Finally , it is also assumed that the insulator is thick enough so that tunneling through the barrier does not occur.

Equations (1) to (3) may be written in normalized form by the use of the following definitions

$$V = qv/kT,$$

$$F = fqL_{Do}/kT,$$

$$X = x/L_{Do} ,$$

$$N(X) = n(x)/n(o),$$

$$J = jL_{Do}/qn(o)D,$$

$$T = t/\tau_R,$$

$$\tag{3}$$

121

with

$$L_{Do} = \left(\kappa T \varepsilon / q^2 n(o)\right)^{1/2}$$

and

$$\tau_R = \varepsilon / q \mu n(o),$$

Where the carrier concentrations are normalized to the mobile carrier concentration at X=0, distances to a Debey length L_{Do}, and times to the relaxation time τ_R appropiate for that value.

The normalized equations are then given by

$$J = NF + \frac{\partial N}{\partial X} + \frac{\partial F}{\partial T} \tag{4}$$

and

$$\frac{\partial F}{\partial X} = - N. \tag{5}$$

We are concerned in this paper with the response of the thick M_1-I-M_2 barrier to an applied time-varying current, in particular, we are considering the situation where a small incremental sinusoidal current J_A is super-imposed upon the steady-state bias J_D. Since the total current is a function of time only we have that the total instantaneous current is give by

$$J = J_D + J_A e^{j\Omega T} \tag{6}$$

Where Ω is the normalized angular frequency. The normalized carrier concentration $N(X,T)$ may accordingly be written as

$$N(X,T) = N_D(X) + N_A(X) e^{j\Omega T} \tag{7}$$

and

$$F(X,T) = F_D(X) + F_A(X) e^{j\Omega T}. \tag{8}$$

Finally, combining equations (4) to (8) and separating the d.c. solution from the incremental a.c. solution yield the following set of equations

$$J_D = N_D F_D + \frac{dN_D}{dX}, \tag{9}$$

$$\frac{dF_D}{dX} = - N_D, \tag{10}$$

$$J_A = F_D N_A + \frac{dN_A}{dX} + (N_D + j\Omega) F_A , \tag{11}$$

and

$$\frac{dF_A}{dX} = - N_A. \tag{12}$$

Using the principle of current continuity together with equations (9) and (10) we obtain a second order nonlinear differential equation with variable coefficients

$$\frac{d^2 N_D}{dX^2} + \frac{1}{N_D}\left(J_D - \frac{dN_D}{dX}\right)\frac{dN_D}{dX} - N_D^2 = 0. \tag{13}$$

This equation has been discused at great length in previous publications[10,11] and has to be used to calculate the quiescent operating condition for a given d.c. current density J_D in the calculation of the a.c. characteristics. In an similar fashion, combining Eqs. (11), where the product term $F_A N_A$ has been neglected, and (12) gives a second order linear differential equation with variable complex coefficients of the complex function N_A

$$\frac{d^2 N_A}{dX^2} + A\frac{dN_A}{dX} - BN_A + \cdot CJ_A = 0 \tag{14}$$

with

$$A \equiv \frac{1}{N_D}\left[\left(J_D - \frac{dN_D}{dX}\right) - \left(\frac{N_D}{N_D + j\Omega}\right)\frac{dN_D}{dX}\right], \tag{15}$$

$$B \equiv 2N_D + j\Omega + \frac{1}{N_D}\left(\frac{1}{N_D + j\Omega}\right)\left(J_D - \frac{dN_D}{dX}\right)\frac{dN_D}{dX}, \tag{16}$$

and

$$C \equiv \left(\frac{1}{N_D + j\Omega}\right)\frac{dN_D}{dX} \tag{17}$$

Eq. (14) may be separated into its real and imaginary parts using the following notation,

$$
\begin{aligned}
N_A(X) &= N_{AR}(X) + jN_{AI}(X), \\
A(X) &= A_R(X) + jA_I(X), \\
B(X) &= B_R(X) + jB_I(X), \\
C(X) &= C_R(X) + jC_I(X),
\end{aligned} \tag{18}
$$

yielding the following two ordinary coupled differential equations with variable coefficients

$$\frac{d^2 N_{AR}}{dX^2} + A_R\frac{dN_{AR}}{dX} - A_I\frac{dN_{AI}}{dX} - B_R N_{AR} + B_I N_{AI} + C_R J_A = 0 \tag{19}$$

and,

$$\frac{d^2 N_{AI}}{dX^2} + A_R\frac{dN_{AI}}{dX} + A_I\frac{dN_{AR}}{dX} - B_R N_{AI} - B_I N_{AR} + C_I J_A = 0. \tag{20}$$

The six variable coefficients are easily calculated from Eqs. (15) to (17) in terms of the quiescent operating condition and the normalized angular frequency of the incremental sinusoidal signal

$$A_R = \frac{1}{N_D}\left[J_D - \left(\frac{2N_D^2 + \Omega^2}{N_D^2 + \Omega^2}\right)\frac{dN_D}{dX}\right], \tag{21}$$

$$A_I = \frac{\Omega}{N_D^2 + \Omega^2} \frac{dN_D}{dX} , \tag{22}$$

$$B_R = 2 N_D + \left(\frac{1}{N_D^2 + \Omega^2} \right) \left(J_D - \frac{dN_D}{dX} \right) \frac{dN_D}{dX} , \tag{23}$$

$$B_I = \Omega \left[1 - \left(\frac{1}{N_D^2 + \Omega^2} \right) \left(J_D - \frac{dN_D}{dX} \right) \left(\frac{1}{N_D} \frac{dN_D}{dX} \right) \right] , \tag{24}$$

$$C_R = \left(\frac{N_D}{N_D^2 + \Omega^2} \right) \frac{dN_D}{dX} , \tag{25}$$

and,

$$C_I = - \Omega \left(\frac{1}{N_D^2 + \Omega^2} \right) \frac{dN_D}{dX} . \tag{26}$$

In conjuntion with Eq. (13), one has to solve numerically Eqs. (19) and (20) for given values of the d.c. current density J_D, the incremental a.c. current density J_A, and the angular frequency Ω. The d.c. Eq. (13) is solved with the usual boundary conditions of constant carrier concentration at the two metal-insulator interfaces, namely, $N(0)$ and $N(L)$ where L is the normalized device length. The real and imaginary parts of the a.c. incremental carrier density may be calculated from Eqs. (19) and (20) postulating as boundary conditions

$$N_{AR}(0) = N_{AI}(0) = N_{AR}(L) = N_{AI}(L) = 0. \tag{27}$$

Knowing the values of $N_{AR}(X)$ and $N_{AI}(X)$ for all X, the real and imaginary parts of the a.c. incremental voltage drop over the device are easily calculated. Finally, the real and imaginary parts of the incremental impedance are obtained from

$$Re \ Z(\Omega) = \frac{V_{AR}}{J_A} , \tag{28}$$

and

$$Im \ Z \ (\Omega) = \frac{V_{AR}}{J_A} . \tag{29}$$

The outlined general approach for the calculation of the incremental impedance of a M_1-I-M_2 barrier is complex. Therefore, if the impedance may be represented by an equivalent circuit consisting of linear, passive, reciprocal, lumped network components then it is more simple and advantageous to calculate immediately the driving-point impedance function $Z(s)$ as described next.

2. Driving-Point Impedance Calculation

Let us assume that the incremental equivalent circuit of the thick

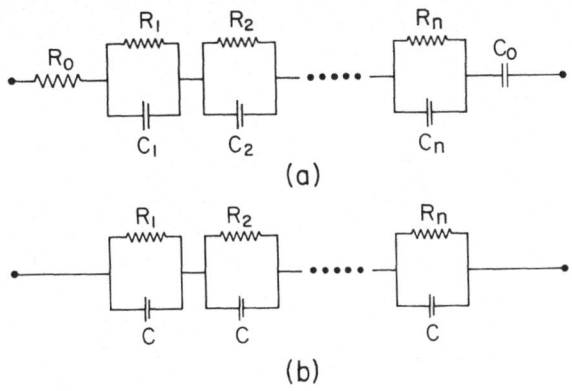

Fig. 1.a) RC Foster network, b) Distributed RC network after[17].

M_1-I-M_2 barrier may be modeled by means of a linear, passive, reciprocal, lumped and time-invariant network. Then, $Z(s)$, with s the complex variable, denotes the driving-point impedance function of the M_1-I-M_2 device defined as the ratio of the response transform to an excitation transform with all initial conditions equal to zero. Also, from Callarotti's discussion[17] we may conclude that the equivalent network of the M_1-I-M_2 barrier is given by a Foster network. The general form of a network composed of a series connection of RC Foster components is given in Fig. 1a, while Fig. 1b shows the proposed network by Callarotti. It can be shown[18] that the driving - point impedance function of RC or, for that matter, RL Foster networks are real for real values of s, and have poles and zeros which are simple and interlace each other along the negative real axis. The RC and RL Foster network functions differ from each other with respect to the sign of their derivative with respect to σ, the real value of s, i.e.

$$\frac{dZ(\sigma)}{d\sigma} < 0 \quad \text{for a RC network,}$$

and,

$$\frac{dZ(\sigma)}{d\sigma} > 0 \quad \text{for a RL network.}$$

As a consequence, plots of $Z(\sigma)$ vs σ show very distinctive patterns. Figs. 2 and 3 give typical plots for the RC and RL driving-point impedance functions $Z_{RC}(\sigma)$ and $Z_{LC}(\sigma)$ vs σ respectively. From such graphs, the poles and zeros can be found, and subsequently the driving-point impedance functions may be readly determined. The equivalent circuit model can be obtained by means of standard methods of networks synthesis.

The calculation of the driving-point impedance is straightforward. It suffices to replace $j\Omega$ in Eqs. (14) to (17) by σ. $N_A(X)$ is now a real function and the numerical calculation is simplified. In doing so, however, a computational instability is introduced due to the term $1/(N_D+\sigma)$ which appears in Eqs. (14) to (17). To circunvent this difficulty we combine equations (11) and (12) to obtain a second order linear differential equation in terms of the incremental electric field F_A,

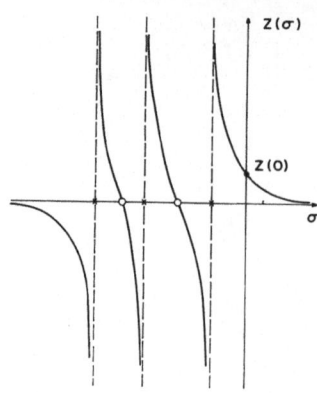

Fig. 2. Typical $Z_{RC}(\sigma)$

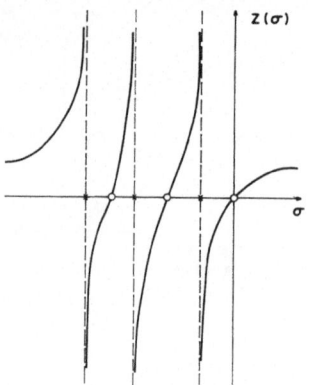

Fig. 3. Typical $Z_{RL}(\sigma)$

$$\frac{d^2 F_A}{dX^2} - \frac{1}{N_D} \left(J_D - \frac{dN_D}{dX} \right) \frac{dF_A}{dX} + \left(N_D + \sigma \right) F_A - J_A = 0 \quad (30)$$

This equation is readly solved numerically using the shooting method with the following boundary conditions

$$\frac{dF_A}{dX} \Big|_{X = 0} = N_A(0) = 0, \quad (31)$$

and

$$\frac{dF_A}{dX} \Big|_{X = L} = N_A(L) = 0. \quad (32)$$

Knowing $F_A(X)$, the incremental voltage drop V_A over the device may be computed and, finally, the driving-point impedance function

$$Z(\sigma) = \frac{V_A}{J_A} \quad (33)$$

is obtained.

We have used this latter procedure for the calculation of the incremental impedance of a thick metal-insulator-metal barrier with $N_D(0) = 1$, $N_D(L) = 100$, and $L = (1/L_{Do}) = 1$. In Figs. (4) to (7) the plots of $Z(\sigma)$ vs σ are depicted for J=0, J=1, J=10 and J=100 respectively. From these figures it is seen that the incremental impedance of the M_1-I-M_2 barrier is apparently a one-pole impedance function with zeros at $\pm$ infinite for a given bias condition. Also it is observed that $(dZ(\sigma)/d\sigma) < 0$. Such an impedance function is easily synthesized by a single RC Foster cell with

$$Z(s) = \frac{R}{RCs+1} . \quad (34)$$

For s=0, Z(0)=R, and is the normalized incremental resistance that can be

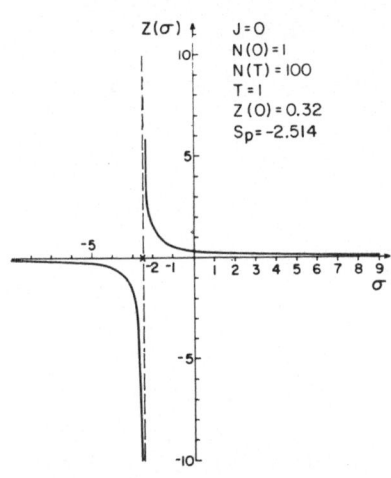

Fig.4. $Z(\sigma)$ for M_1-I-M_2, $N(0)=1$, $N(L)=100$, $J=0$.

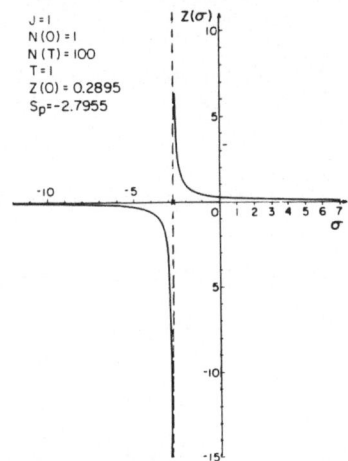

Fig.5. $Z(\sigma)$ for M_1-I-M_2, $N(0)=1$, $N(L)=100$, $J=1$.

calculated directly from the d.c. current-voltage characteristic. From the knowledge of R and of the pole location which is given theoretically by

$$\sigma_{pt} = - \frac{1}{RC} , \tag{35}$$

where, $$Z(\sigma_{pt}) = \pm \infty ,$$

the value of the capacitance can be calculated. Of course, the exact theoretical pole location can never be computed numerically. With the computational method used, it results that it is easier to approach the discontinuity from the left side than from the right side and that there exists a range of values of σ for which no numerical solution is available. For example, for the case of J=0, on the left side of the discontinuity the nearest value is given by $\sigma_1 = -2.514$ with $Z(\sigma_1) \cong - 2238$ while on the right side $\sigma_r = -2.385$ with $Z(\sigma_r) \cong + 6.23$. Thus, the exact pole location is bounded by

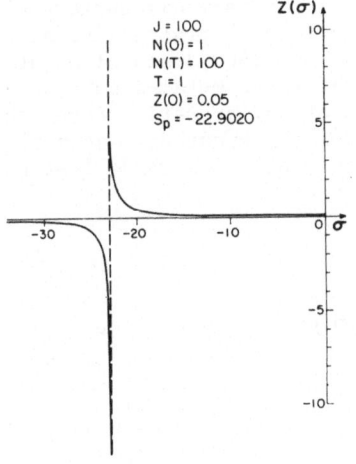

Fig. 6. $Z(\sigma)$ for M_1-I-M_2, $N(0)=1$, $N(L)-100$, $J=10$.

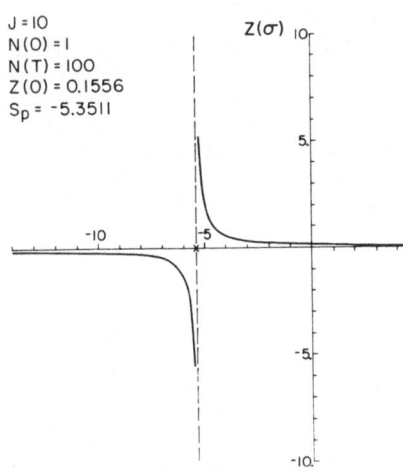

Fig. 7. $Z(\sigma)$ for M_1-I-M_2, $N(0)=1$, $N(L)=100$, $J=100$.

$$-2.514 < \sigma_{pt} < -2.385.$$

A best fit to the numerical calculated plots was found to occur for the value of the pole calculated as follows

$$\frac{1}{\tau_p} = \sigma_p = \frac{Z(\sigma_1)\sigma_1}{Z(0)-Z(\sigma_1)} \tag{36}$$

yielding the following impedance function,

$$Z(\sigma) = \frac{Z(0)}{\tau_p \sigma + 1} \tag{37}$$

A reasonable fit is established near the pole but large errors (>10%) do occur for large values of $|\sigma|$. It is not clear at the present time if these errors are due to computational errors, or if the observed discrepancy is indicative of the fact that there might exist another pole for more negative values of σ, such that a series connection of two RC Foster cells would be necessary in order to represent adequately the MIM device at higher frequencies.

CONCLUSIONS

We have discussed a general approach for the calculation of the incremental impedance from the single-carrier drift-diffusion equation for a two-terminal device. An alternative method has been given for the case of a two-terminal device that may be modeled by means of a linear, passive, reciprocal, lumped and time-invariant network. If the latter possibility is assumed to be valid for the thick M_1-I-M_2 barrier, then the driving-point impedance function may be calculated. The obtained results indicate that the MIM barrier may be modeled satisfactorily in the low frequency range for all bias conditions as a one pole RC impedance function.

ACKNOWLEDGEMENT

It is with fondness that we think back on the many discussions with Professor Heinz Henisch concerning contact problems on crystalline and amorphous materials. It is during one of these numerous discussions at our laboratory that he brought to our attention the thick metal-insulator-metal barrier problem. During all these years of collaborative research we have had the privilege to know Professor Heinz Henisch as a scientist with a big mind, as a teacher with a great sense of humor, and, last but not least, as a friend with a big hart.

REFERENCES

1. M.A. Lampert and P. Mark, "Current Injection in Solids, " Academic Press, New York (1970).
2. N. F. Mott and R. W. Gurney, "Electronic Processes in Ionic Crystals, " 2nd ed., Oxford University, Oxford (1948).
3. H. Y. Fan, "Theory of Rectification of an Insulating Layer," Phys. Rev. 74:1505 (1955).
4. W. Schockley and R.C. Prim, "Space-Charge Limited Emission in Semiconductors," Phys. Rev. 90:753 (1953).

5. G. T. Wright, "Mechanisms of Space-Charge Limited Current in Solids," Solid-State ELectron. 2:165 (1961).
6. S. M. Skinner, "Diffusion, Static Charges and the Conduction of Electricity in Nonmetallic Solids by a Single Charge Carrier. II Solution of the Rectifier Equations for Insulating Layers," J. Appl. Phys. 25:509 (1955).
7. J. Lindmayer, J. Reynolds and C. Wrigley, "One-Carrier-Space-Charge Limited Current in Solids," J. Appl. Phys. 34:809 (1963).
8. M. A. Lampert and F.E. Edelman, "Theory of One-Carrier Space-Charge Limited Currents Including Diffusion and Trapping," J. Appl. Phys. 35:2971 (1964).
9. J. S. Bonham and D.H. Jarvis, "A New Approach to Space-Charge Limited Conduction Theory," Aust. J. Chem. 30:705 (1977).
10. H. K. Henisch, J.C. Manifacier, R.C. Callarotti and P.E. Schmidt, "Space-Charge Controlled Conduction in Thick Metal-Insulator-Metal Barriers," J. Appl. Phys. 51:3790 (1980).
11. P. E. Schmidt, M. Octavio and R.C. Callarotti, "Single-Carrier Space-Charge Controlled Conduction," J. Appl. Phys. 53:4996 (1982).
12. J. Shao and G.T. Wright, "Characteristics of the Space-Charge Limited Dielectric Diode at Very High Frequencies," Solid-State Electronics, 3:291 (1961).
13. H. Yoshimura, "Space-Charge Limited and Emitter Current Limited Injections in Space-Charge Regions of Semiconductors," IEEE Trans. Electron Devices, ED. 1:414 (1964).
14. F. B. Llewellyn, "Vacuum Tube Electronics at Ultra-High Frequencies," Bell Syst. Tech. J. 13:59 (1934).
15. F. B. Llewellyn, "Operation of Ultra-High Frequency Vaccum Tubes," Bell Syst. Tech. J. 14:632 (1935).
16. G. T. Wright, "Small-Signal Characteristics of Semiconductor Punch-Through Injection and Transit-Time Diodes," Solid-State Electronics, 16:903 (1973).
17. R. C. Callarotti, "Proper Capacitance Modeling for Devices with Distributed Space-Charge," preceding paper.
18. S. Seshu and N. Balabanian, "Linear Network Analysis," John Wiley, New York, Third Printing, Ch. 9 (1964).

DIELECTRIC BEHAVIOUR OF AMORPHOUS THIN FILMS[*]

K.B.R. Varma, K.J. Rao and C.N.R. Rao[**]

Materials Research Laboratory
Indian Institute of Science
Bangalore 560 012 India

ABSTRACT

Dielectric behaviour and ultramicrostructures of r.f. sputtered films are examined in as-sputtered and crystallized conditions. $LiNbO_3$, $K_3Li_2Nb_5O_{15}$, $KTaO_3$ and KH_2PO_4 amorphous films with high dielectric constants and exhibit ferroelectric-like dielectric anomalies. This apparently universal dielectric behaviour of thin ionic films is examined using a cluster model. Anharmonicity in the vibrations of particles in the tissue region of the films is considered to be the possible origin of the dielectric behaviour.

INTRODUCTION

Although thin films of many dielectrics are x-ray amorphous, they often exhibit high dielectric constants and dielectric anomalies akin to those of crystalline ferroelectric materials. This curious phenomenon of amorphous materials exhibiting ferroelectric-like transitions was first noted by Glass et al (1977) in the case of glassy films of $LiNbO_3$. We reported (Varma et al, 1985) sometime ago a similar behaviour in r.f. sputtered films of $LiNbO_3$. The anomaly occurs just before the glass transition temperature. The glass transition in such materials occurs close to the crystallization temperature and consequently only crystallization is observed in most thermal experiments.

Lines (1977) has viewed the incidence of ferroelectricity in glassy $LiNbO_3$ as due to the presence of dielectrically soft units. R.f. sputtered thin films indeed possess an ultramicrostructure composed of ordered regions which may be described as microcrystallites of about 20Å size. We have provided evidence (Parthasarathy et al 1983; Rao 1984) elsewhere for the presence of intermediate range order in all amorphous materials and especially in those which are ionic. Ferroelectric-like behaviour is unlikely to be directly related to the ultramicrocrystalline structure of thin films. It seems however possible that the disordered material which holds the microcrystallites together in amorphous thin films and glasses is responsible for the dielectric anomaly. The anomaly in amorphous $LiNbO_3$, it may be noted, occurs in the same temperature range in the glassy as well as the r.f. sputtered films (Varma et al 1985).

---

[*]Contribution No.83 from Materials Research Laboratory

[**]to whom all correspondence should be addressed

It is interesting to ponder whether the dielectric anomalies would occur in all ionic materials prepared in the amorphous state. If so, we should expect ferroelectric like anomalies in thin films of materials just before their crystallization or glass transition temperatures even when the material is not ferroelectric in the crystalline state. Furthermore, the temperature at which such anomalies occur should have no relation to the Curie temperatures, if any of the crystalline materials. In this paper we have examined the above possibilities.

EXPERIMENTAL

The following compounds have been examined in this work in thin film form: $LiNbO_3$(LN), $K_3Li_2Nb_5O_{15}$(KLN), $KTaO_3$(KT), KH_2PO_4(KDP) and KCl. Crystalline KLN powder was investigated in pellet form in order to compare with literature reports.

LN, KLN and KT were prepared starting from analar grade K_2CO_3, Li_2CO_3, Nb_2O_5 and Ta_2O_5. Two different KLN samples were prepared containing 50 and 53.1 mole % Nb_2O_5. A mixture containing alkali carbonates and Nb_2O_5 in required proportions was ground to a fine powder and heated in a platinum crucible to 1000K and soaked at that temperature for about 8 hours. The product was cooled, again ground to a fine powder, compacted and reheated slowly to 1250K and maintained at that temperature for about 48 hours. It was then cooled to room temperature. Powder x-ray diffraction of the product was examined and the formation of KLN was confirmed. Pellets of the two different KLN compositions were obtained by sintering the compacted material once again at 1250K for about 48 hours. LN and KT were prepared similarly starting from Li_2CO_3, Nb_2O_5, K_2CO_3 and Ta_2O_5.

The materials were r.f. sputtered on to a quartz substrate held at room temperature using MRC (Materials Research Corporation, USA) r.f. sputtering unit model SEM-8622 provided with a 13.56 MHz r.f. generator capable of providing a maximum output power level of 1.5 KW. Sputtering was performed in argon plasma at a constant pressure of 4×10^{-3} torr. Oxygen was continuously leaked into the chamber in a controlled way while sputtering. The structure of the sputtered films were examined using a high resolution transmission electron microscope, model JEOL, JEM 200 CX. The operating voltage of the system was 200 KV with a primary magnification of $3-5 \times 10^5$.

Gold electrodes were deposited using masks prior to sputtering the films required for dielectric constant measurements and a planar sample configuration was adopted for measurements. Capacitance was measured as a function of both temperature (300-1000K) and a frequency (1-100 KHz) with a signal strength of 5V rms using a three terminal General Radio Capacitance Bridge (USA) coupled externally with GR 1316 oscillator and GR 1232A tuned amplifier and a null detector. The stray capacitance introduced by leads etc., was minimized by reducing the lengths of the connecting leads and by providing adequate electrical shielding. Dielectric constants of films were evaluated after correcting for the residual stray capacitance.

RESULTS

The Temperature variation of the logarithm of dielectric constant of the sputtered films of $LiNbO_3$(LN) is shown in Fig.1. A similar plot of the data of Glass et al(1977) obtained using LN glass is also presented for comparison. HREM micrographs of r.f. sputtered LN films in as-sputtered and crystallized states are shown in Fig.2; in the inset this figure electron diffraction patterns are provided.

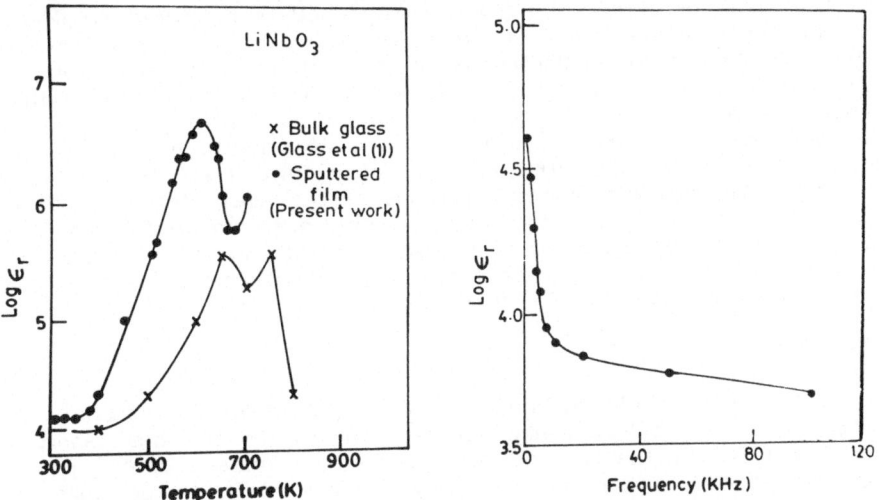

Fig.1. Temperature variation of the dielectric constant of r.f. sputtered LiNbO$_3$ (LN) film and LN glass at 1 KHz; data on glass is from Glass et al (1977).

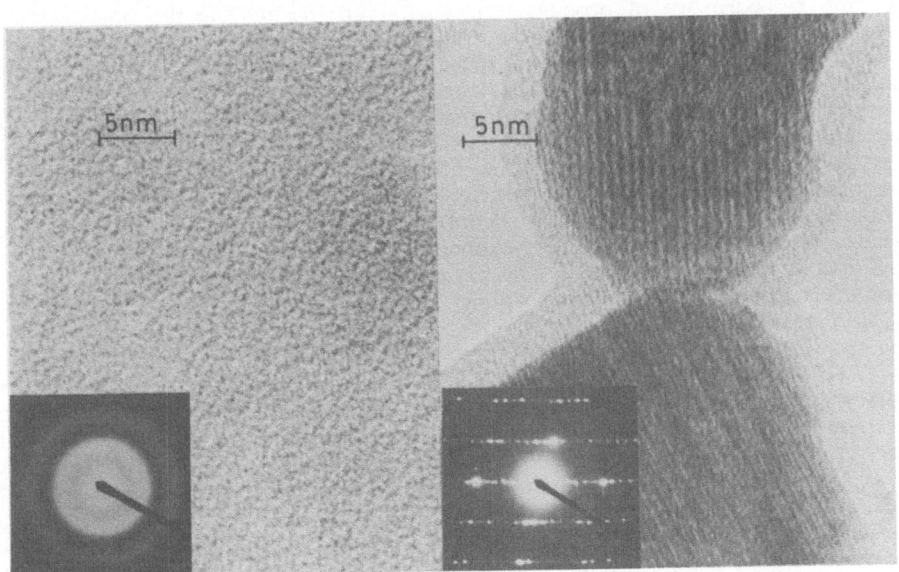

Fig.2. High resolution electron micrograph of as-sputtered (left) and crystallized (right) LN films. The corresponding diffraction pattern is shown in the inset.

Dielectric constants of KLN films and pellets measured at 1KHz are shown in Fig.3. In the inset of Fig.3, we show the frequency dependance of the logarithm of the dielectric constant. High resolution electron micrograph of as-sputtered film of KLN is shown in Fig.4. along with that of the heat treated sample (for about 12 hours around 870K); diffraction patterns for respective cases are also shown in this figure. In Fig.5., we show a semilog plot of the dielectric constant of a film of KT against temperature. High resolution electron micrographs of the as-sputtered KT film and the film annealed at 1070K (for about 8 hours) are compared in Fig.6. along with the respective diffraction patterns. The dielectric behaviour of the sputtered KDP film is shown in Fig.7 and high-resolution electron micrographs of the amorphous film in Fig.8.

DISCUSSION

R.f. sputtered LN films exhibit very high dielectric constants. More interestingly, a ferroelectric-like anomaly similar to the one reported by Glass et al (1977) in the case of fast quenched LN glass is observed in these films. The anomaly occurs at a slightly lower ($\backsim$50K) temperature in the film, but the dielectric constant is higher by an order of magnitude. The temperature of dielectric anomaly however occurs at much lower temperature ($\backsim$600K) than the ferroelectric-paraelectric transition (T_c) (Subba Rao, 1974) of crystalline LN (1483K).

KLN films exhibit dielectric anomalies similar to crystalline KLN pellets, but at temperatures lower (by about 200-300K) than in the crystalline materials. The dielectric constants of the thin films are at least three orders of magnitude higher than the crystalline materials investigated in the form of pellets. The dielectric anomalies found by us in the pellets are comparable to those observed by Scott et al (1970) although both the peak temperatures and the magnitudes of the dielectric constants were higher in our study. It is possible that some amount of Nb_2O_5 was lost from our samples at the high temperature of preparation causing an increase in the peak temperature and the peak dielectric constant; another possibility is that the sintered pellets in our experiments were more porous giving rise to higher surface contribution to polarization. The more interesting observation is that in amorphous KLN films, the dielectric anomaly occurs at a higher temperature than in the crystalline sample. The rise in dielectric constant is also spread over a wider temperature range and the peak occurs at the crystallization temperature. Occurence of crystallization is also confirmed by electron microscopy (Fig.4). This may be contrasted with the behaviour of LN discussed earlier where the dielectric anomaly in the film occured at a temperature much lower than the T_c in crystalline LN. However, in thin films of both LN and KLN, the dielectric constant peak occurs just before the crystallization transition.

KT has a very low T_c (about 2K) (Subba Rao, 1974) and has no other dielectric anomaly. KT films however exhibit a rather intense dielectric anomaly (Fig.5) which peaks around 475K and above this temperature the film starts crystallizing rapidly. With the progress of crystallization, there is a decrease in the dielectric constant as expected. Here again the thin film of KT is quite amorphous (Fig.4). That the KT film exhibits a dielectric anomaly at a temperature 470K higher than the T_c of crystalline KT excludes the microcrystallite origin of the dielectric anomaly in the amorphous films.

The dielectric behaviour of KH_2PO_4 films further confirms the above inference. While crystalline KDP itself has a fairly low transition temperature (123K) (Subba Rao, 1974) sputtered films exhibit a rather high dielectric constant peak around 430K (Fig.7). Above this temperature, the films crystallize (Fig.8). The composition of the sputtered film was, however, somewhat different from that of the starting KDP as inferred from lattice para-

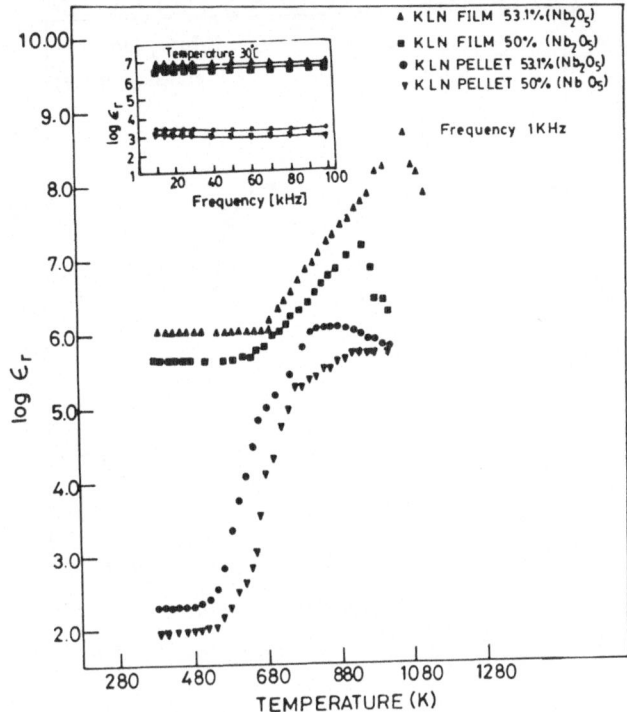

Fig.3. Semilog plots of the dielectric constants of KLN films and pellets against temperature at 1KHz frequency. Inset shows the frequency dependance of the dielectric constant.

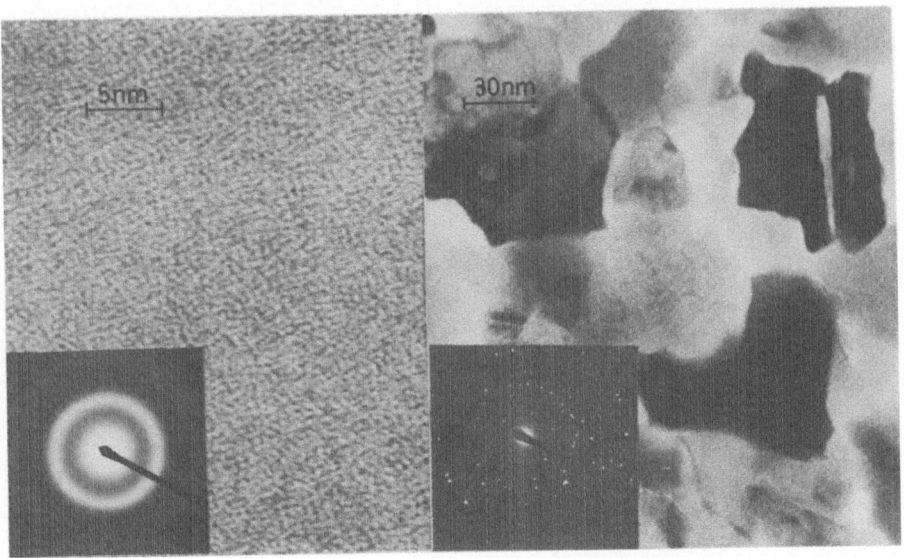

Fig.4. High resolution electron micrographs as sputtered (left) and annealed (12hrs at 870K) (right) films of KLN (50% Nb_2O_5). The corresponding diffraction patterns are shown in the inset.

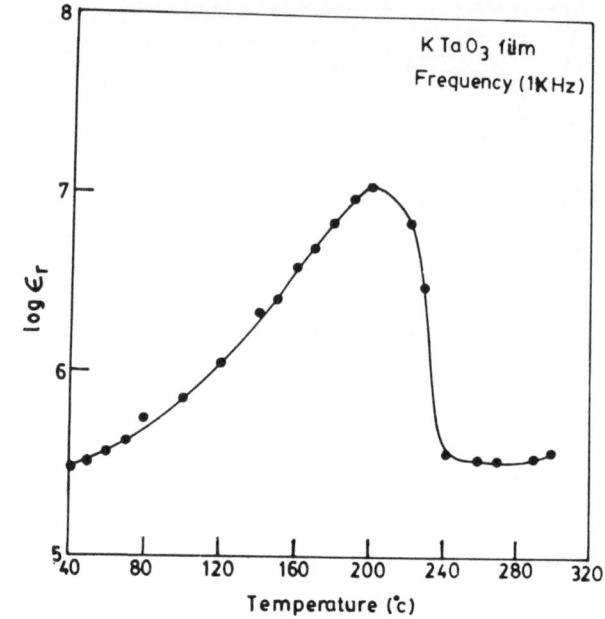

Fig.5. Semilog plot of the dielectric constant of KT film against temperature (1KHz frequency).

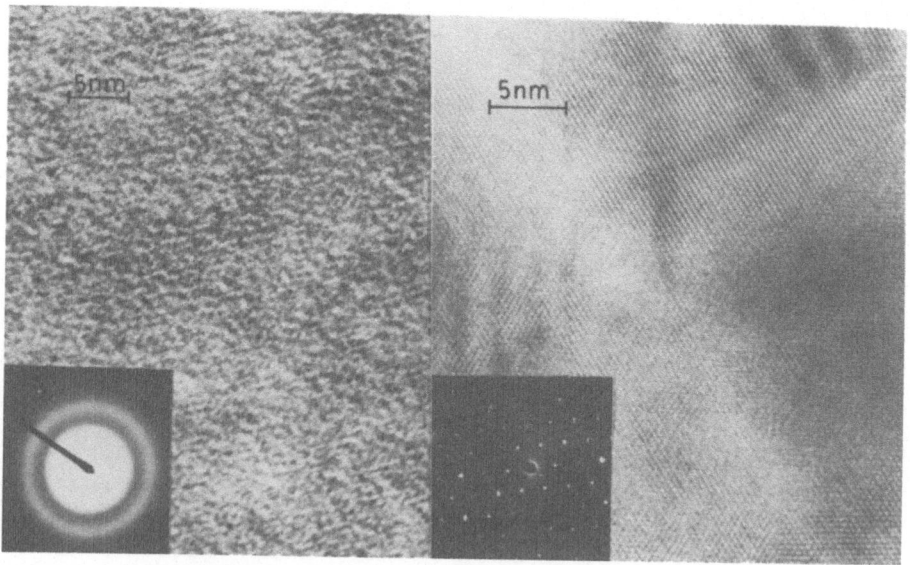

Fig.6. High resolution electron micrographs of as-sputtered (left) and annealed (8hrs at 1070K) (right) KT Films with respective diffraction pattern shown as insets.

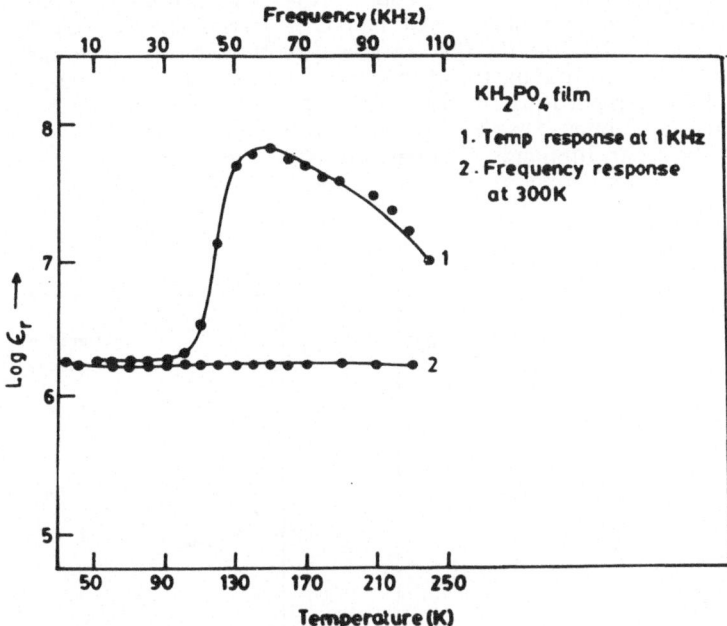

Fig.7. Semilog Plot of the dielectric constant of KDP Film against temperature (1KHz frequency) and the frequency response of dielectric
constant at 300 K.

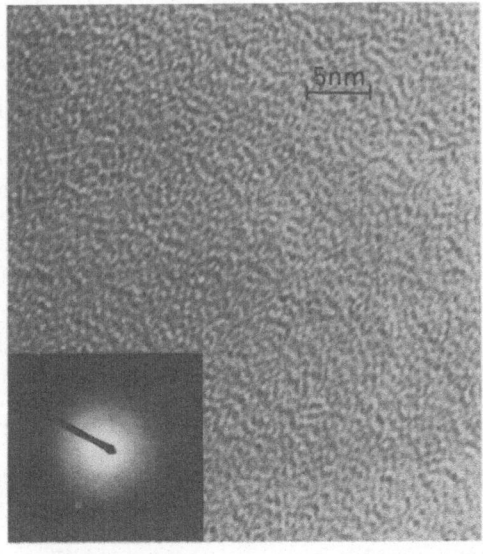

Fig.8. High resolution electron micrograph of KDP Film. Electron diffraction pattern is shown in the inset.

meters. It is interesting that amorphous KCl films also exhibit high dielectric constants. Although crystalline KCl is not a ferroelectric, KCl films exhibit a dielectric anomaly (Fig.9). The anomaly itself is not so prominent although the dielectric constants of the films were quite high. The anomaly occured over a small temperature range. As-sputtered films of KCl were found to be quite amorphous by HREM and to rapidly crystallize following the anomaly; these preliminary findings on KCl are being further investiga-

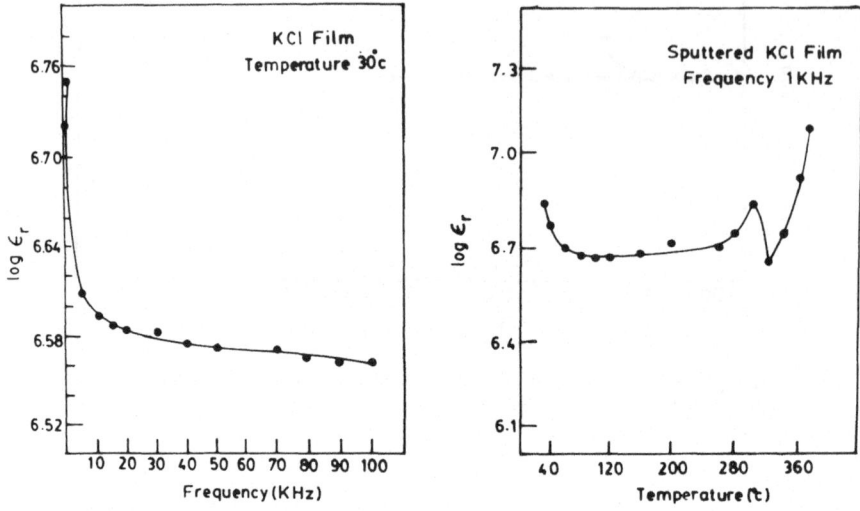

Fig.9. Semilog plot of the dielectric constant of KCl thin film as a function of temperature and the frequency response of dielectric constant at 300 K.

High Dielectric Constants of Thin Films

Thin films of LN, KLN, KT, KDP, KCl all possess very high dielectric constants and exhibit ferroelectric-like anomalies prior to their crystallization temperatures. All these films possess ultramicrostructures which can be considered as due to intermediate range order of around 15-20Å dimension. The dielectric anomaly in thin films bear no relation to the dielectric properties of parent crystalline materials. The higher magnitudes of dielectric anomalies appear to be associated with the presence of more polarizable oxy-anions and highly polarizing cations.

The above observations raise a vital question regarding the origin of the large dielectric constants of thin films. The latter could be related to the presence of a fine mosaic of more' and 'less' spatially ordered regions in the thin films. The more ordered regions in the limit correspond to microcrystallites and characterize the intermediate range order in amorphous films. The less ordered regions would correspond to completely disordered amorphous regions in the thin films (Rao and Rao, 1982; Parthasarathy et al, 1983, Rao, 1984). It seems likely that all amorphous materials possess such ultramicroheterogeneous structures (Rao, 1986). Because of the rather small spatial dimensions, the ordered regions or the microcrystallites do not produce crystalline x-ray diffraction patterns. Further-

138

more, they are likely to possess highly symmetrical nonpolar structures for thermodynamic reasons. Microcrystallites or regions of intermediate-range order do not therefore produce the necessary dipole moments or spontaneous polarization which can give rise to high dielectric constants as is the case of crystalline ferroelectrics. The nature of the disordered region is interesting; it would be slightly lower in density since it has to accommodate spatial disorder. This is the case with random close packing of identical spheres as well. When such a disordered material is subjected to an electrical field, it behaves like an electrolytic capacitor. The large assembly of such micro-capacitors in the thin films can give rise to high dielectric constants which therefore become characteristic of amorphous thin films.

Dielectric Anomaly and the Cluster Model

The temperature variation of the dielectric constant of ultramicrocrystalline thin films can be understood on the basis of the cluster model. The cluster model emphasizes the anharmonicity of vibrations of the constituent particles. The temperature regime over which the amorphous films are stable (till it terminates in a transition) is dominated by the anharmonic processes in the disordered or the so called 'tissue', region. The tissue region itself grows in volume at the expense of the clusters as the temperature increases and is characterized by extremely anharmonic vibrational energy potentials. Because of the anharmonicity, the vibrational energy levels coalesce and the excitation energies decrease in such coalesced wells at higher vibrational levels. Correspondingly, the vibrational amplitudes also increase. We have shown elsewhere that the configurational properties of glasses emerge as a function of temperature when the higher vibrational levels get populated (Rao, 1984).

We attribute the configurational properties of the amorphous films of LN, KLN, KT, KDP and KCl to the vibrational properties of the cations in the tissue region. A vibrational energy scheme has been suggested in the model according to which the successive energy separations in the vibrational energy manifold are related as $\Delta E_{n} = \Delta E_{1}/n$, ΔE_{1} is itself equal to $h\omega_{1}$ where ω_{1} is the vibrational frequency and n indicates the nth vibrational level.

$$\Delta E_{n}/\Delta E_{1} = \omega_{n}/\omega_{1} = 1/n \qquad \text{------------ (1)}$$

As the temperature is increased, the ions are excited to higher vibrational energy levels from where their further excitation require less energy and correspondingly the vibrational mode 'softens'. In the cluster model, it has been suggested that the mode which thus 'softens' in discrete steps finally evolves into a diffusive mode towards the glass transition. Therefore, in amorphous LN, for example, the mode is associated with Li^{+} ion motion and may be considered as a 'soft mode' of the amorphous phase. The model thus provides a plausible soft mode basis for understanding the observed dielectric anomaly. The mode, however, evolves into a translation much before it softens out entirely at the glass transition and therefore the familiar dielectric catastrophy does not occur. Or, the mode may also suddenly get hardened when interrupted by crystallization. Such unusual mode softening has been shown to explain the observed temperature variation of Lamb-Mossbauer factors in (Bharathi et al 1983) in glasses. The same soft mode description can now be used to understand the dielectric behaviour of amorphous films.

An approximate estimate of the effect of the mode softening upon the dielectric constants of the amorphous material may be made as follows:

139

If we take the frequency of the softening mode associated with the cation motion as ω_{TO} and use Lyddane - Sachs - Teller (LST) relation(Kittel, 1968),

$$\frac{\varepsilon_o}{\varepsilon_\infty} = \frac{\omega_{LO}^2}{\omega_{TO}^2} \qquad \dots \qquad (2)$$

where ω_{LO} and ω_{TO} are the transverse and longitudinal optical mode frequencies respectively. We have

$$\varepsilon_o = \varepsilon_\infty \omega_{LO}^2 / \omega_{TO}^2$$

According to the above equation, the dielectric constant should rapidly rise as ω_{TO} decreases. We assume that ω_{TO} is the ω_i of the cluster model. ω_i is 'softer' for larger value of the index i of the vibrational level.

$$\varepsilon_o = \Sigma_i \, \varepsilon_{i,0} \qquad \dots \qquad (3)$$

Hence

$$\varepsilon_o = \text{Constant} \left[\Sigma_i \, \frac{f_i}{\omega_i^2} \cdot \frac{V_{tissue}}{V_{total}} \right] \qquad \dots \qquad (4)$$

where constant $= (\omega_{LO}^2 \cdot \varepsilon_\infty)$, f_i is the fraction of the relevant ions in the ith vibrational state and V_{tissue}/V_{total} is the fractional volume of the tissue. However, as the temperature increases the cluster volume decrease and the rigidity of the matrix provided by the clusters also vanishes. Hence the 'soft mode' we are considering becomes increasingly highly localized loosing its significance as 'phonon' mode. We further assume that the number of effective (phonon like) soft modes is directly proportional to the volume fraction of clusters and obtain the expression for ε_o as

$$\varepsilon_o = \text{Constant} \left[\Sigma \, \frac{f_i}{\omega_i^2} \cdot \frac{V_{tissue}}{V_{total}} \cdot \frac{V_{cluster}}{V_{total}} \right] \qquad \dots \quad (5)$$

The product $(V_{cluster} \times V_{tissue})$ which becomes zero at T_g ensures incidence of maximum in the dielectric constant irrespective of the detailed description of f_i while $\Sigma_i f_i /\omega_i^2$ is a continuously increasing function.

The significant point we make here is that the dielectric anomaly of amorphous materials most likely originates from the tissue and not from the microcrystalline cluster regions. Not only the high dielectric constant but the occurance of a ferroelectric-like transition close to the glass or crystallization transition is also explained by the cluster-tissue description of thin films. The equillibrium of cluster and tissue in the thin films is not disturbed after annealing the films unless the transition temperature is traversed. Thus thin ionic dielectric films particularly those composed of large polarizable anions and small cations can be easily adopted in opto-electronics requiring high dielectric constant materials.

CONCLUSIONS

Amorphous films of ionic dielectrics exhibit large dielectric constants and ferroelectric-like dielectric anomalies in the region of crystallization. The dielectric anomaly in the crystalline state and in the amorphous films are not directly related. Amorphous films appear to possess cluster-tissue texture in as-sputtered condition and crystallizes when heated. Tissue regions behave like electrolytic capacitors giving rise to high dielectric constants. The dielectric anomaly is a consequence of anharmonicity of particle vibrations in the tissue and is understood qualitatively using cluster model.

REFERENCES

Bharathi S, Parthasarathy R, Rao K J and Rao C N R, 1983, Solid State Communications, $\underline{46}$, 457.

Glass A M, Lines M E, Nassau K and Shiever J W, 1977, Appl. Phys.Letters $\underline{31}$, 249.

Kittel C, 1968 Introduction to Solid State Physics John Wiley, NY.

Lines M E, 1977, Phys.Rev. $\underline{15}$, 388.

Parthasarathy R, Rao K J and Rao C N R, 1983, Chem.Soc.Rev., $\underline{12}$, 361.

Rao K J, 1984, Proc.Indian Acad.Sci. (Chem.Sci) $\underline{93}$, 389.

Rao K J, 1986, Proc.Indian Natl.Sci.Acad. 52A, 176.

Rao K J and Rao C N R, 1982, Mat.Res.Bull $\underline{17}$, 1337.

Scott B A, Giess E A, Olson B L, Burns G, Smith A W and O'Kane D F, 1970, Mat.Res.Bull. $\underline{5}$, 47.

Subba Rao E C, 1974, in Solid State Chemistry Ed. by C N R Rao, h.11, Marcell Dekker, N Y.

Varma K B R, Harshavardhan K S, Rao K J and Rao C N R, 1985, Mat.Res.Bull. $\underline{20}$, 315.

A PERSONAL ADVENTURE IN STEREOCHEMISTRY, LOCAL ORDER, AND DEFECTS: MODELS FOR ROOM-TEMPERATURE SUPERCONDUCTIVITY

Stanford R. Ovshinsky

Energy Conversion Devices, Inc.
1675 West Maple Road, Troy, MI 48084

ABSTRACT

The relationship between electronic conduction and stereochemical local bonding arrangements is illustrated by four examples: the unusual high conductance ON-state of the Ovonic Threshold Switch resulting from nonequilibrium excitation processes, the high conductivity connected with the oxidation states of the Ovitron, the chemical modification of electronic transport in amorphous semiconductors and insulators, and the high temperature copper oxide ceramic superconductors. We discuss the effect of fluorination on raising the superconducting transition temperature to $T_c=155K$ or higher, on eliminating oxygen diffusion and on orienting the crystallites in the YBaCuO superconductors.

I.   INTRODUCTION

Heinz Henisch's contributions to semiconductor physics have had a great effect on me. Many years before I met him 20 years ago, I often used his well-known book "Rectifying Semiconductor Contacts."[1] His important contributions to the amorphous field are many; we cite just a few [2-6] which did much to establish the electronic nature of Ovonic threshold switching.[7,8]

Heinz's work is intertwined with ours both in the past and currently with his suggestion that the Ovonic Threshold Switch (OTS) could be a superconducting device.[9] In this paper, I discuss some of our work in superconductivity and propose two models, one of which speculates on such a possibility for the OTS and the other which, as is my wont, does not emphasize the crystalline nature of the new high-temperature superconducting materials but rather structural chemical concepts that I have developed for amorphicity, such as normal structural bonding (NSB), deviant electronic configurations (DECs), and total interactive environment (TIE).[10-13] I describe how the unusual electronic properties of multi-elemental materials can be related to their steric chemistry in such a way that high-temperature superconductors ensue.

The lack of theoretical understanding of the funda-mental mechanism of the new high-temperature supercon-ductors has led to many theories and models. It is an exhilarating time because there is a freshness of thinking that has swept the physics world and new ideas are stimulating much work. I believe that there is not just one but several mechanisms for achieving high-temperature superconductors.

In this paper I shall discuss two quite different systems in which superconductivity can occur. The first may be termed equilibrium superconductors. These are the conventional superconductors as well as the new high-T_c superconducting materials. In these, the superconducting state below T_c is a state of lowest free energy. The second system may be called nonequilibrium superconduc-tors. I propose that in these, the conducting charge plasma is created by external means and may become superconducting below T_c. This could be accomplished by a large gate voltage producing a metallic state in the inversion layer of a semiconductor, or by a strong light excitation producing, above the Mott insulator-metal criterion, a metallic electron-hole plasma and electron-hole liquid droplets in crystalline semiconductors, or by strong double injection which yields a conducting filament in an amorphous semiconductor OTS. Even though the nonmetal metal transition has been established in these nonequilibrium systems, the transition to the superconducting state has not yet been observed with certainty. This theme will therefore be highly speculative.

I will begin by discussing the arguments for and against the possibility that the conducting On-state of an OTS is carried by a superconducting filament. I then proceed by explaining the stereochemical relationships which characterize the unique properties of amorphous materials and which make chemical modification of the conductivity of amorphous semiconductors possible. I close by describing a stereochemical model for the high-T_c superconductivity observed in the Yttrium-Barium-Copper-Oxide ceramic materials, which I believe has general applicability to other mixed valence systems. Finally, the evidence for raising the critical temperature T_c to 155K and higher by fluorination will be presented.

II. CONDUCTING STATE OF THE OVONIC THRESHOLD SWITCH

Figure 1 shows the current-voltage characteristic of an Ovonic Threshold Switch (OTS). This device is typically

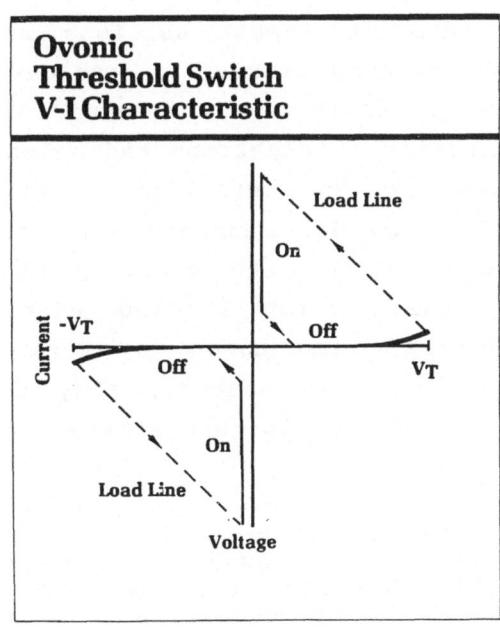

Ovonic Threshold Switch V-I Characteristic

Fig. 1

made of a 0.5μm thick multicomponent non-crystalline chalcogenide semiconductor contacted on both sides by non-alloying contacts having a diameter of a few microns.[7] Without an applied voltage, the OTS is in the high resistance or OFF-state. At the threshold voltage, V_T, the OTS switches to the ON-state which will be examined in this section. The switching occurs so fast that the speed was never accurately established because of limitations due to external inductances and capacitances. The switching time is less than a picosecond which makes this the fastest room-temperature switch. The Josephson junction later reached this speed at liquid helium temperatures.

There are several factors suggesting superconductivity as the origin of the ON-state. Henisch [9] reviews "In the ON-state, the resistance is independent of electrode area, signifying that the current flows in a filament that is ordinarily much smaller, e.g. $10^{-8} cm^2$, than conventional electrode areas. Accordingly, the current densities are enormous.[*] In the ON-state, the resistance is almost independent of system thickness, signifying that the potential drop is close to the electrodes, not in the bulk of the semiconducting material. The electric field free bulk is the region of special interest here."

The conducting ON-state is stable as long as the current exceeds a minimum holding current as indicated in Fig. 1. This conducting state is believed to be established by double injection of electrons and holes from the electrodes. The injected carriers first fill the localized gap states and in so doing smooth out existing potential fluctuations and enhance the mobility of free carriers. Depending on the exciton binding energy, a fraction of the free electrons and holes may form bound excitons. Since such excitons are neutral, they do not promote the conducting state. It is more likely that the

---

[*] In thin-film devices, we at ECD measure a minimum of 2×10^4 amps/cm^2. Our early data suggest that shaped electrodes which focus the electric field in very small areas can produce higher current densities. Threshold values are of the order of 10^5 V/cm.[14]

density of electron-hole pairs exceeds the Mott criterion for the nonmetal-metal transition. We estimate that this concentration is about 10^{19}/cc with carrier effective masses of $0.5m_e$ and a dielectric constant of 12.

As previously mentioned, essentially all of the voltage drop across a threshold switch in the ON-state occurs at the contacts. Such would be the case if the electron-hole plasma in the conducting filament were actually superconducting at room temperature or above. Ovonic threshold switches contain chalcogenides in which the highest unoccupied states form a lone-pair band. In the ON-State, the Fermi level splits into two quasi-Fermi levels with the hole level deep enough for there to be a large hole density in the lone-pair band and similarly for electrons in the antibonding conduction band. If one now invokes a negative effective correlation energy in these materials showing that there is a large, attractive Hubbard interaction, U, between two holes or between two electrons, then such an interaction could lead to superconductivity by Bose condensation of electron pairs, hole pairs, or both, the pairs being bipolaron-like, [15] or of the Bardeen-Cooper-Schrieffer (BCS) type [16] where the unique very strong electron-phonon interaction of these materials could be invoked to provide pairing.

I have pointed out in many papers the physical basis of the negative effective correlation energy in the chalcogenides in relation to the lone pairs and their relevance to the excitation process of the OTS [17-22] and superconductivity.[*] Lone pairs are not only unbonded but can have various bonding configurations such as one- and three-center bonds as well as dative/coordinative bonds [21] and the elegant valence alternation pairs of the Kastner-Adler-Fritzsche (KAF) model.[22] The point that I want to emphasize here is that in the unbonded state the lone-pair configuration with its spin up and spin down can be viewed as a localized Bose particle.

---

* See quote on page 12.

The Bose particles formed by excitation then follow Bose-Einstein statistics, leading to condensation to the ground state at a particular temperature and density of excited particles. In fact, the large number of particles in such an excited state cannot exceed a certain critical value _without_ reaching a saturation condition that would result in Bose-Einstein condensation and superconductivity. The assumed velocity and effective mass of the particles support the model.

In the OTS one can see excitation processes prior to the formation of the filament which involve the generation of carriers from the bulk of the material (see Fig. 2). Only when the filament is formed and the self-regulating plasma confinement results in a high constant current density with a fixed volume and excitation temperature is it possible that the Bose-Einstein conditions can be met.

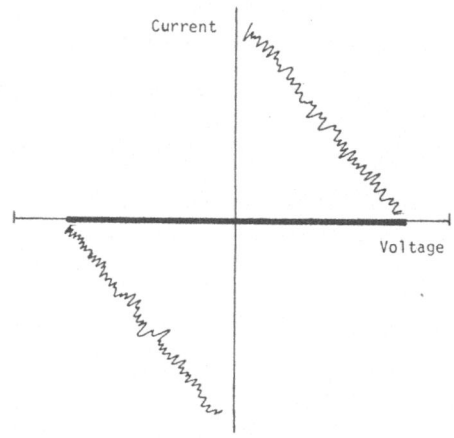

Fig. 2 The Ovonic Threshold Switch in the oscillatory mode.[23]

It is interesting that the actual OTS "ON-state" plasma is not affected by an external magnetic field and that these amorphous chalcogenide materials are normally diamagnetic due to their lone-pair character with compensated spin up, spin down configurations. Of interest to our stereochemical model of the copper oxide superconductors to be discussed later is the commonality of chain structures, crosslinking/bridging atoms and mixed valency and coordination states which I have used to describe the

physics and chemistry of amorphicity.[8] Of passing
interest, the OTS would like to undergo a structural phase
transition but is prevented from doing so by strongly
bonded crosslinking atoms.

The fact that, as a result of their interaction
with the local environment, lone pairs can move without
breaking [24] has led me to a thought problem which can be
outlined as follows: While the energetics do not favor it,
considering the Cohen-Fritzsche-Ovshinsky (CFO) model [25]
and allowing for a spectrum of lone pair-states, some
strongly localized, some weakly so, could it be possible
that under a very high electric field in a very thin
amorphous lone-pair device with a high density of states, a
sufficient number of weakly localized Bose pairs could
tunnel as pairs, providing another device possibility. It
is interesting to consider that the "wingless" OTS [23] can
represent the tunneling mechanism while a thicker version
would favor excitation.

I have considered Little's [26], Ginzburg's [27] and
Bardeen's [28] excitonic models and have the following
thoughts about them. Obviously, the many materials and
configurations tried through the years to achieve their
type of excitonic superconductivity have not been
fruitful. However, if one considers a conducting plasma as
a "surface," replacing the layer postulated by Ginzburg,
and then interfaces and encloses that plasma with an
ionizable medium such as a high-resistance chalcogenide or
other dielectric, then it would seem to me that the
requirements for pairing of their hypothesized excitonic
mechanism might be met.

Some historical reflections before I discuss our
latest work on high T_c superconductors. I was never able
to go so far myself as to publicly consider, as Henisch now
boldly does, that the OTS is a room-temperature superconductor. However, I did consider in the 1960's that the
conducting state was so unusual, for the reasons that I
have discussed above, that at some temperature probably
lower than room-temperature the ON state might go

superconducting. There were several attempts made by others to find superconducting correlations with our devices. One mistake made was that a researcher confused the memory crystalline ON-state with the threshold plasma ON-state and did not find any high-temperature superconductivity, as indeed there was no reason to assume that he would.[29] On the other hand, Fritzsche and Sakai addressed the problem by taking arsenic telluride and subjecting it to high pressure so as to move the chains together and to reach a superconducting state.[30] The results were interesting in that they showed that an amorphous chalcogenide alloy could have a superconducting transition although it was only at 3K. In my view, what was missing were the excitation processes necessary to provide the conditions for high-temperature superconductivity.

The motivating factor for our superconducting work with amorphous materials in the 1960's and 1970's was that I believed that without lattice restrictions there were possible new orbital relationships and interactions necessary to develop high temperature superconductors. This was contrary to the dogma of that time that superconductivity depended upon specific crystal structure.

I showed that amorphous materials could have orders of magnitude higher critical magnetic field capability than had been reported till then and we later were able to get a very respectable zero resistance superconductivity of 9K.[31] J.T. Chen, who spent a sabbatical at ECD, developed with us the technique of the inverse Josephson junction effect that has been so helpful in indirectly measuring high-temperature superconductivity.[32] Strongin at Brookhaven had developed a precursor of the technique.[33]

Why would I have considered the chalcogenides as suitable candidates for superconductors? My chemical reasoning goes back to my work begun in the mid-1950's on designing conductivity changes in rare earth oxides and particularly in transition metal oxides including copper oxides.[8,34] My interest in f- and especially d-orbitals was connected with my interest in disorder where, depending upon the local environment, one could have a variety of electronic configurations that are not available in crystalline materials. Of special interest in this history is my Ovitron device (see Fig. 3) of 1957 where I utilized transition metal ("valve") oxides such as tantalum to achieve an over 14 orders of magnitude drop of resistance

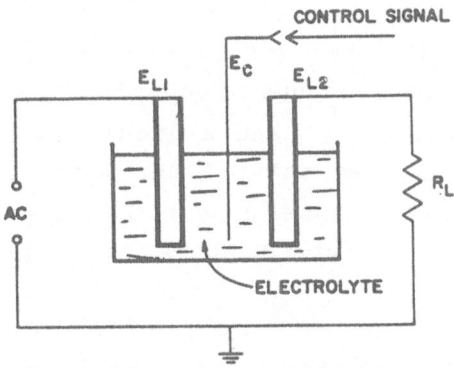

Fig. 3 Schematic diagram of amorphous dielectric film switch and modulator. The load resistor R_L and the amorphous film on the anodized tantalum electrodes E_{L1} and E_{L2} submersed in electrolyte form the load circuit. Current flows through the load circuit if a positive signal is applied to the control electrode E_C. Gain and, under certain conditions, memory are observed when metallic ions influence the blocking properties of the amorphous dielectric films which are shown for AC operation back to back.

in response to a small signal through "...the interaction of the metallic ion with the amorphous film..."[35] The metallic ion was divalent zinc (which can be amphoteric). This device was based upon a thin film of electrochemically formed amorphous oxide of tantalum containing small amounts of other elements such as a halogen. I was uncertain of the oxidation state of the sub-oxide, and the effect of the minor constituents in the electrolyte upon the oxidation state of the tantalum was never fully determined. I felt that the interaction of the zinc orbitals with the tantalum oxide altered the coordination, transformed the valency and therefore the band structure so that the oxide went from a dielectric to a conductor which carried large current densities. This dynamic interaction can be considered to have similarities to chemical modification which I developed in the 1970's [36] and may be related to the mixed valency, charge balance, superconducting mechanism to be discussed.

In the Ovitron device, there had to be a balance of charge as the metal ion interaction altered the oxidation state with a simultaneous reduction action at another electrode. I originally used a liquid electrolyte. It was of great interest to me that if the metal ions were removed from the electrolyte, the unique features were lost. In the late 1950's, I adapted these principles to the solid state.[35]

Relevant to our model for high-temperature supercon-ductivity to be discussed, there was a reservoir of carriers, a chemical "pump" and a simultaneous electron transfer and exchange resulting in a valence transformation utilized for these unique conductivity changes.[8,35]

III. CONDUCTANCE CHANGES BY CHEMICAL MODIFICATION

Unusual conductivity changes intrigued me (I even investigated the conductivity changes of metal liquid ammonia systems), and I set out to show that one could independently control the conductivity in amorphous and disordered semiconductors and in wide band gap dielectrics without altering other important parameters unless desired. I achieved conductivity changes of over 10 orders of magnitude through chemical modification in materials that spanned the Periodic Chart (Fig. 4). In the fully modified state, metallic-like conduction was achieved (Fig. 5).[13,36-38] This chemical modification was reproduced by many prominent scientists such as Kolomiets et al., Davis and Mytilineou, etc.[39-41]

152

The important factor was that I showed that one could transform (chemically modify) what would normally be considered large band gap dielectrics consisting of elements and alloys utilized in ceramics such as silicon carbide and silicon nitride into surprisingly good conductors while still keeping their large band gap intact. Chemical modification was not only counter-intuitive but demonstrated that the conductivity of amorphous chalcogenides could be controlled, contrary to the generally-held view that large changes of conductivity

CHEMICAL MODIFICATION OF AMORPHOUS SEMICONDUCTORS	
Host Material	Active Modifier
Ge Te Se As	Ni, Fe, Mo
As	Ni, W
$B_4 C$	W
Si C	W
Si	Ni, B, C
$Si_3 N_4$	W
BN	W
Te O_2	Ni
Ge	Ni
Si O_2	W
Ge Se_2	Ni
$Se_{95} As_5$	Ni

Fig. 4

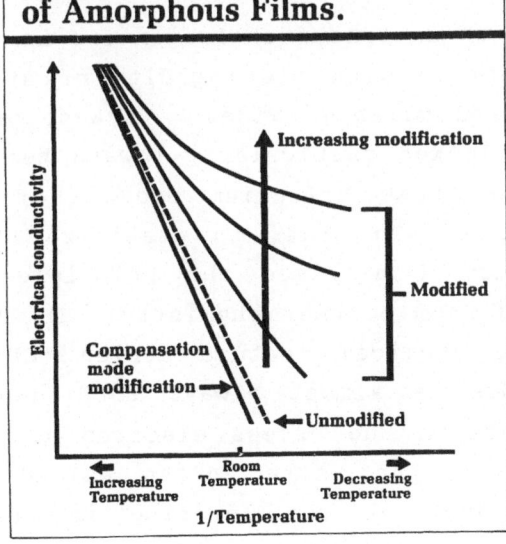

Effect of Ovonic Modification on Electrical Properties of Amorphous Films.

Fig. 5

such as those obtained by doping of crystalline materials were not possible. I showed that new orbital relationships could be established in a material in a stable, nonequilibrium manner so as to control its electronic properties without resorting to substitutional dopants and dramatically, emphasized the importance that I attributed to stereochemistry in the solid state. I feel that chemical modification is relevant to the new high-temperature ceramic superconductors. The d-orbital interactions of yttrium and like elements with copper oxide can also affect pairing and provide increased conductivity.

What was of interest to me was not only the unusual transformation of conductivity achieved by chemical modification but that, despite the transition elements used, the materials could be diamagnetic. When one considers that deposited films have a high density of states, such as dangling bonds and configurations that would normally be paramagnetic, one can assume that the transition metal d orbitals are being utilized for pairing. In terms of the chemistry of transition metals, maximum use is made of all available particles, that is, holes and electrons, interacting so that pairing becomes the preferred mode as in low spin crystal field splitting where various states are close in energy to one another.

I decided to work with amorphous chalcogenides in 1960 because of my dissatisfaction with the rigidity of structures involved in the mixed valence oxides. Because of my biological motivation, I looked particularly toward helical and chain structures. I chose tellurium because of its helical/chain configuration. In order to stabilize these one-dimensional structures, I utilized the principles of bridging and crosslinking from polymer chemistry. In 1977, I wrote for a polymeric chemical lecture series "Charge transfer in semiconductors is almost always forbidden by the Coulomb repulsion between the excess electron and the other electrons already on that site, often called the correlation energy. Because of the lowering in energy resulting from the bond-breaking described above, in chalcogenide glasses the effective correlation energy is negative. (The resulting effective attraction between electrons is also analogous to superconductivity.) This leads to many of the unique properties of chalcogenide glasses, including the OTS switching phenomena."[42,43] I was particularly interested in the anisotropic qualities of the chain-like structures and how to utilize them electronically while structurally transforming them into configurations of higher dimensionality.

The valency and coordination transformations in the Ovitron and my other mixed valency oxide work are a bridge between my early work and our model for high-temperature superconductors which follows. Much of my past and present work depends upon the structural chemistry of mixed valence materials and on the orbital interactions in one-, two- and three-dimensional space of elements that under ordinary conditions would not "see," that is, interact with, each other in space and energy. With its dependence on a rigid monolithic lattice, conventional crystalline thinking seems to me to be of little use to the new high-temperature superconductors.[24,44]

As I wrote in Mott's Festschrift in 1985, "I have stated ... that the Rosetta Stone of amorphous materials is the understanding of the relationship between the normal structural bonding (NSB) which characterizes the great majority of atoms and is responsible for the cohesiveness of the amorphous solid and the deviant electronic configurations (DECs) that control the transport properties and provide the active chemical sites of the material. Tying these two together is the concept of the total interactive environment (TIE) ... which takes into account the special nature of various local, chemical, topological, and electronic interactions in amorphous solids. We can write a new language of materials if we make use of this new alphabet. We need not be limited to the old dogmas of homogeneity and equilibrium chemistries. We have a new world of nonequilibrium chemistry and varying topological structures. <u>We can deliberately create combinations and geometries in which even the local short-range order can vary subtly or drastically from one part of the material to</u>

the other. Such new structural chemistry again produces new electronic phenomena and new sites for chemical activity. Amorphous materials with unique cluster configurations and those containing crystalline inclusions and layers are also part of the spectrum of engineered materials discussed here. Such designed materials have far-reaching applications. We can synthesize and engineer materials where we mismatch and compensate atoms without the problems of mismatching lattices. We can carry this further through the use of layering and compositional modulation.... In fact, heterogeneity then becomes a welcome tool rather than a scare word."[24] (Emphasis in original.)

IV. SUPERCONDUCTIVITY AT HIGH TEMPERATURES

Starting with my basic structural chemical concept explaining high-temperature superconducting materials, my collaborators and I developed a model which describes the fundamental mechanism that produces the high-temperature superconducting state. It is very specific, testable, self consistent and has predictive properties. We have published a paper presenting it in greater depth and detail [45] and portions of it appear here. In the present paper I shall advance some new thoughts which seek to dispel the mysteries of sheets and chains and show how as long as one has interacting and transformable oxidation states of +2's and +3's in desired geometric relationships, whether a material is lanthanum- yttrium- or bismuth-based copper oxide, [46] it can show high-temperature superconductivity. The physics follows directly from the stereochemically-determined structures, defects and local and total interactive environments just as it does in amorphous materials.

As seen in Fig. 6, the crystal structure of orthorhombic $YBa_2Cu_3O_7$ has copper atoms occupying two distinct sites. Copper is present both in dimpled two dimensional copper-oxygen sheets located between the yttrium and barium layers, and in one dimensional copper-oxygen-copper chains, formed by the ordered oxygen vacancies located between the two barium layers. It is exceptional that not only are the chains created by the vacancies but that the barium, or a like element such as strontium, does not only provide proper overall charge compensation, specific valency and size but is in place to stabilize what might otherwise be an unstable chain structure. I suggest that barium, which has such an affinity to oxygen that it is classically used as a getter, interacts with the lone pairs of oxygen assuring the unique chain structure.

I feel that the ferroelectric characteristics of these superconducting materials are also relevant. As Von Hippel pointed out years ago in connection with conventional ferro-

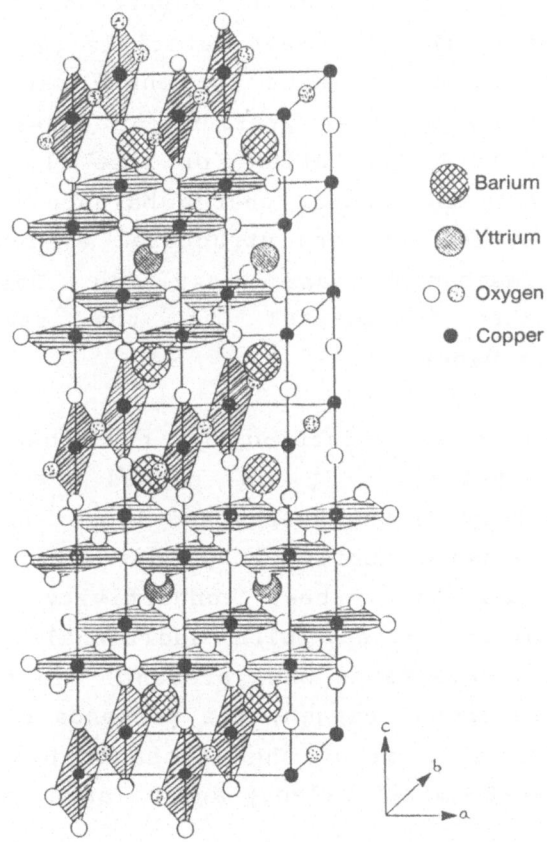

Fig. 6 The crystal structure of $YBa_2Cu_3O_7$. In this figure cross-hatched planes are used to clearly delineate the "sheet" and "chain" copper sites. Although the "sheets" are shown as planar, the Cu-O-Cu bonds are actually puckered with a bond angle of 165°. Bridging oxygen atoms in the "chains" are shaded in the figure for ease in identification.

electric materials, "Furthermore, the role of the barium ions is more important than that of a simple charge compensator and glue... These cations add their own dipole system to the overall balance, alter the size, shape, and deformability of the oxygen octahedra, and may tilt their mutual orientation."[47]

Of the three copper atoms in the unit cell, two are present in sheet sites and one is in a chain site. The chemical bonding and, particularly, the valence in these two sites are different. It is this difference of valency that permits the transfer of electrons and holes as well as setting up the conditions for antiferromagnetic coupling which is necessary for high T_c superconductivity. Such mixed valency can exist in sheet materials (e.g., lanthanum or bismuth) and in mixed sheet and chain materials (e.g., yttrium). Elements such as yttrium (or other rare earths of the same valency) and barium (or like elements such as strontium) provide the proper charge balance to assure the mixed valence system and are structural elements as well. The pairing mechanism and resulting Bose particle condensation are fundamental to the stereochemistry detailed in this paper.

Anisotropic superconduction in the sheets has been considered a mystery. It is caused by insufficient transfer sites between the sheets and limits the attainment of the optimum superconductivity by inhibiting the required Bose particle density. Sheet conductivity is basically dependent on mixed valency with additional contributions from d-orbital interactions when present. In the lanthanum sheet materials, excess oxygen, the presence of defects, or the substitution of some of the lanthanum by strontium or barium can generate mixed valency and balance charges.

To achieve the density requirements of the Bose-Einstein formula, there must be short-range connecting sites joining the sheets/chains, promoting the carrier exchange between them. They provide the needed change from anisotropic to isotropic geometries and conductivities. Lone pairs, because of their spatial flexibility, are helpful in this respect.[8] Materials with chain structures, with their dimensionally critical orbital overlaps, are formed by the ordered oxygen vacancies. Without these vacancies, the ~ 90K zero resistance superconductivity could not be achieved in the YBaCuO systems. Twinning and other structural distortions can generate new localized chemical configurations and connections which can represent new phases, some of which may be the hard-to-achieve, very high-temperature ones (over 125K).

The interactions described below are essential for high-temperature superconductivity in the YBaCuO systems. The copper in the dimpled two-dimensional copper-oxygen sheets has a nominal valency of two and in the chain sites a nominal valency of three. It is important to note that the copper oxygen bond angle for the layers is 165° and for the chain it is exactly 180°. Copper III is an <u>unusual</u> state and the 180° angle <u>exceptional</u>. The majority of copper II spins in the sheets is coupled antiferromagnetically, resulting in no local magnetic moment. Antiferromagnetic coupling comes about through an oxygen intermediary by the superexchange process. The antiparallel spin alignment utilizes the oxygen lone-pair p orbitals to form the bond and is favored by the large copper-oxygen-copper bond angle. Some of my favorite configurations, lone pairs and coordinate bonds, play an important role through the oxygen-mediated superexchange process. Pairing strength depends upon bond angles and bond lengths. Undesirable spatial separations weaken the antiferromagnetic coupling and disrupt the continuity of the valence transformation process and affects the density of Bose particles.

The copper atoms sit at the center of a planar array of oxygen atoms, forming copper-oxygen-copper chains between the "tunnels" created by the ordered oxygen vacancies which are present in the orthorhombic structure. Our identification of the valence state of these atoms as spinless copper III with a d^8 configuration is consistent with observed coordination and bond angles, and is dictated by our previous assignment of copper II to the copper atoms in the sheets. The four nearest neighbor oxygen atoms lie roughly in the sheet with the fifth, the apex oxygen atom, showing a Jahn-Teller distortion [48] and located at a larger bond length below the copper along the crystallographic c axis. In my view, this apex oxygen transfer configuration is critical.

The basis for high-temperature superconductivity is the establishment of simultaneous valence transformation processes involving two atoms at a time <u>throughout</u> the unit cell. This is accomplished by the interactions between the sheets and the chains through the aforementioned "pump"-like electron transfer mechanism of the pyramidal linking structure. As electrons are transferred up and down from the adjacent sheets to the copper-oxygen-copper chains, a mixed copper II, copper III valence state is formed in the sheets and the chains. Each electron transferred from the sheet leaves behind a copper III and converts one of the copper III atoms on the chain to copper II. This dynamic process is the heart of the new high-temperature superconducting materials.

This process does not produce localized copper II atoms on the chains, nor does it leave copper II atoms with unpaired spins on the sheets since this would produce local magnetic moments. Rather the d orbital holes present in this mixed valence state are delocalized, giving rise to the observed Pauli-Landau temperature independent paramagnetic susceptibility. These delocalized holes, however, still interact with one another via the oxygen atom intermediaries and, at a particular temperature, two spins on alternate sides of a bridging oxygen atom can be in a favorable position, because of the large bond angle, to interact through the superexchange process to produce an anti-parallel spin pair. The spin pairs which are thus formed can also <u>migrate</u>, now in a <u>bound state</u>, in what can be considered to be a <u>simultaneous valence transformation</u> process involving two atoms at a time.

<u>The superexchange coupled spin pairs on the chains and</u>

sheets are mobile, strongly bound, spinless composite particles which obey Bose statistics. At any given concentration of these spin pairs, there will exist a Bose condensation temperature at which a transition to a superconducting ground state will occur. There are, therefore, three important temperatures to be considered. The first two are the spin pairing temperatures for the d orbital holes in the sheets and in the chains respectively and the third is the Bose condensation temperature. The superconducting state is achieved at the lowest of these temperatures. It is possible, of course, that the lowest temperature could be the Bose condensation temperature. This novel situation would have dramatic consequences for the electrical properties of the normal state, to say the least, resulting in the occurrence of charge transport through uncondensed spin pairs! It is possible that under certain conditions such uncondensed charge carrier pairs can participate in "normal" conduction.

We can obtain an estimate of the Bose condensation temperature for the model system in the following way. The volume occupied by the wavefunction of a Bose particle can be approximated by a cube with dimensions equal to the particle's de Broglie wavelength. The de Broglie wavelength, in turn, is determined by the particle's momentum and, therefore, its thermal energy. This temperature dependent Bose particle interaction volume therefore increases with decreasing temperature. When the interaction volume grows to become equal to the volume available per Bose particle in the system, the Bose particles interact so as to bring about condensation. Using this approach, calculations show that the greatest density of Bose particles will occur when two-thirds of the chain copper atoms have copper II valence. This will, of course, also require that each sheet is one-third empty. One obtains, for this occupancy, a density of carrier pairs, $n=2.9\times10^{21}cm^{-3}$, or roughly one pair per 6 copper atoms. This carrier density is in close agreement with the free carrier density which is measured [49] above T_c by Hall Effect in $YBa_2Cu_3O_7$. Bose condensation temperatures exceeding room temperature are possible.

Although it is difficult to obtain a realistic estimate of the spin pairing temperatures in terms of this simple model, it is clear that, because of the increased orbital overlap between copper d electrons and oxygen p electrons which occurs in the 180° bond angles along the chains, we would expect spin pairs to be more strongly bound through superexchange in these structures than in the sheets. Support for this is obtained from recent nuclear spin lattice relaxation experiments on the yttrium-barium-copper-oxygen system reported by Warren et al. [50] which clearly indicate the presence of two distinct pairing energies, with substantially larger energies for quasiparticle formation (pair breaking) in the chains than in the sheets.

There is unnecessary confusion as to the role of sheets and chains. The same antiferromagnetic spin pairing mechanism is operative in both. The sheets can contain a large number of +3 configurations interacting with +2's. The mixed valence copper system is established in the sheets either by doping with strontium or barium or through oxygen excess. The chains are unusual stereochemical means of achieving a +3 valency to add to the overall valence transformation possible throughout the sheet-chain apex molecular system.

The above concepts show the importance of local dimensionality and total interactive environment. The spatial-energetic relationships, reflected in bond angles, bond lengths and antiferromagnetic pairing strengths, are important to the mechanisms of both pair formation and Bose particle condensation. I emphasize here my personal view that one must establish the necessary density for Bose condensation, and that this requires not sheets alone but a mechanism transforming two dimensionality into three dimensionality--that is, intersheet connections which establish the necessary containment for the Bose particles.

Mixed valency can be provided by various elements, for example, the amount of oxygen in bismuth materials maintains the +2/+3 ratio of valence states proposed in our model. However, the bismuth materials lack sufficient valence transformation linkages between the sheets. This explains why even with more sheets than the 1,2,3 materials, they do not have optimal zero resistance superconductivity. However, the bond angles in the bismuth sheets appear to be closer to the 180° ideal, and therefore the pair breaking energy would be higher.

The superexchange interaction energy, which produces pairing, becomes the dominant factor as the temperature-dependent lattice vibrations, which destroy pairing, grow smaller with decreasing temperature. At T_c, the spin pairs form and condense into the superconducting Bose state. Large spacings between the sheets and difficult communication between them are antagonistic to high-temperature superconductivity. The high conductivity of anisotropic sheets is misleading since the sheets may contain the proper number of particles for high-temperature superconductivity but lack the proper density to fit the basically isotropic formula.

In our work in high-temperature superconducting materials, I chose fluorine to make a new alloy in which it could be a factor in the control of charge and valency, increase carriers available for superconductivity, make for stronger bonding and be involved not only in the chain configurations but also, under certain conditions, in the bridging apex and sheets. I felt that it would promote stability in a material notorious for its weak oxygen bonding. The results were dramatic. We reported 155-168K zero resistance superconductivity in fluorinated copper oxide ceramics.[51] These results have been confirmed by several groups, for example, in China, [52,53] Taiwan [54,55] and Sweden.[56] Despite our reports [51,57-59] and their confirmation, one can still find statements that there are no confirmed reports of superconductivity above 125K. Figure 7 shows a fluorinated sample (T_c=154K) made by a plasma process [59] rather than the solid phase reaction [51] that we had previously reported.

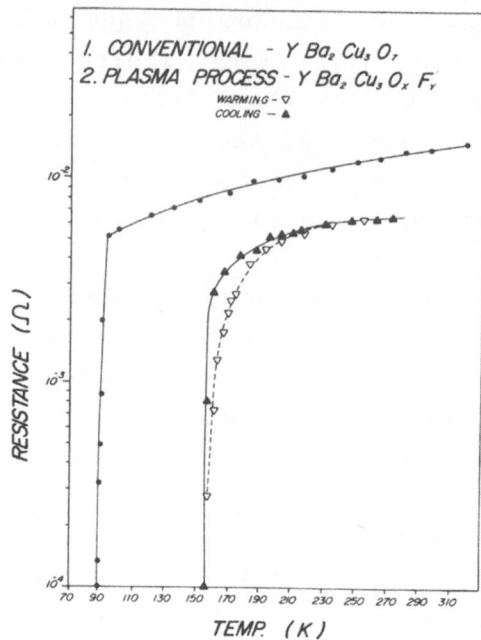

Fig. 7. Resistance vs. temperature plot showing 154K zero resistance transition of a microwave-treated YBaCuOF sample.

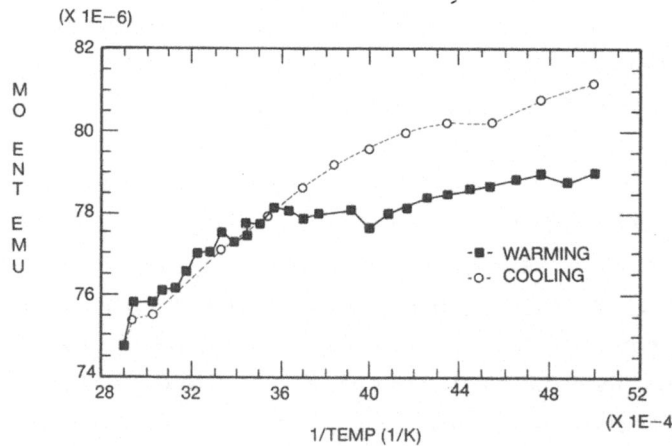

Fig. 8 Magnetic moment vs. 1/T for a YBaCuOF sample. Measurements made with use of 40G field. Data for warming after zero-field cooling are indicated by ■ and data from cooling with field applied are indicated by o.

What is equally exciting is our observation of diamagnetic signals and flux trapping at temperatures as high as 305K, [60] indicating that there are higher than room-temperature phases in our fluorinated materials (see Fig. 8). Supporting this magnetic data is Fig. 9 where we were able to show evidence by conductivity measurements of the existence of a phase exhibiting superconductivity onset above room temperature. It is important to point out that the resistivity of this sample <u>above</u> T_c is four times lower than that of copper![57]

In the YBaCuO material, the role of the chains is that they uniquely provide +3's, add three dimensionality to the system and are in communication with the sheets through the apical structure in such a manner that valence transforma-

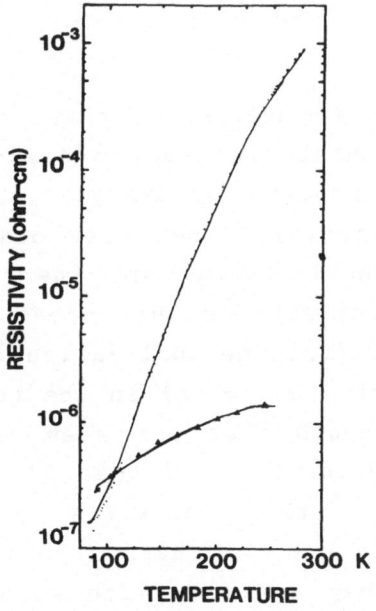

Fig. 9   The logarithm of the average of YBaCuOF sample. Average resistivity was calculated on the assumption of uniform current density. Resistivity was found to follow T^n, where n = 8.3. The ideal resistivity of pure copper is also plotted (triangles).

tion can take place throughout the three configurations making up the molecular structure. They demonstrate the unique 180° bridging bond angle between the copper and the oxygen, and the oxygen-vacancy-generated and barium-stabilized quasi-one-dimensional chain structure which had been considered not to be possible.(61) The chain-sheet dichotomy is a false one since it is the mixed valence molecular structure of which the chain is a component which becomes superconducting.

Crystal structures are relevant to the extent that they geometrically provide maximum particle availability and pair interaction. Whether materials are tetragonal or orthorhombic is not basic but what is important is how the atoms are related to each other in three-dimensional space. They must have not only a mixed valency but also must be able to communicate and interact three dimensionally in space and energy to optimally meet the Bose condensation criteria. Local configurational spacing is important since it determines pairing energy. Overall spacing involved in the molecular structure is important to density. There is much room for improving high-temperature superconductors.

It is important to understand that the short coherence length is a fundamental clue to the described mechanism of high-temperature superconductivity. That the new superconducting materials are not ordinary metals is apparent from consideration of the weak temperature dependence of conductivity in the normal state. We do not see either the electronic delocalization and long mean-free path that one expects in a metal in the normal state or the long interaction length that one sees in normal BCS-type metallic superconductors. I suggest that this is due to the fact that the carriers responsible for superconductivity do not originate from the same source as the metal, i.e., they are very much more localized coming from the mixed valency transfer mechanism. These materials emphasize the small dimensional, localized, tight binding antiferromagnetic interactions that make for the strongest Bose formation pairing energy. There are two spatially, and therefore energetically, controlled, volumetric configurations--the first reflecting the local antiferromagnetic couplings; the second, a larger but still constricted container of Bose particles whose density, and therefore condensation to the superconducting state, is dependent upon their small mean-free path and low mobility. As in semiconductors, unwanted defects can act as recombination centers and even scatter the Bose particles.[*]

The high-temperature ceramic superconductors with their reliance upon connectivity, varied short and intermediate range order, their relatively low mobility and their process dependency are reminiscent of amorphous and disordered materials more than they are of the conventional crystalline materials. Subtle structural relaxations can play a significant role in determining T_c through coupling strength and volumetric changes.

One should learn from miracles. The attempt to continue making superconducting ceramic materials in the conventional way is basically flawed from a materials and mechanism viewpoint. God did not just point his finger at the yttrium-barium-copper oxide crystal structure and say "I have done my work." It is up to us to synthetically design new materials--laying them down almost atomically, layer by layer, as we have done in our x-ray mirrors and even our amorphous "superlattices."[8]

It is easy to be misled by Fig. 6, for it does not take into account unwanted defects, sheet and chain breaks, vacancies, the effect of subtle structural relaxations, bond reconstructions, twinning, interfaces, and the changes of charge that are involved with these factors, all of which are process related and in my view can be controlled or eliminated by the methods of synthesis described above.

---

[*] Based upon this concept, I will discuss in another paper the fundamental differences in mechanism between BCS superconductors and the new high-temperature ceramic superconductors.[62]

Superconductivity exists at and above room temperature. It is difficult to produce such superconducting materials in substantial amounts by present methods. The first transistor illustrates the problems and opportunities of the historical process of creating new devices (Fig. 10). The surface, chemical and structural problems involved in this primitive device are now all but forgotten. I believe that the superconductivity concepts and model outlined above are a guide for our technological and scientific advances. There is a naivete in thinking that the happenstance of the technique of making a bulk T_c = 95K material is relevant to the task of achieving higher T_c materials needed for the next great step forward.

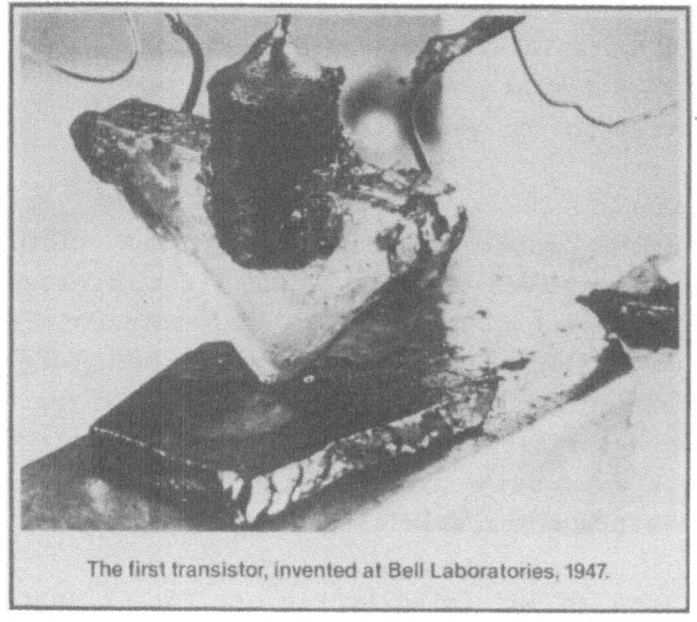

The first transistor, invented at Bell Laboratories, 1947.

Fig. 10

We have made four important contributions through the use of fluorine--the first, utilizing fluorine to make a new high-temperature superconducting material results in the highest confirmed <u>zero resistance</u> superconducting temperature,[51,57] albeit the material is multiphasic with a small volume fraction of the very high-temperature phases (such phases have fortuitously transposed and/or juxtaposed atoms in favorable configurations so that the sheets have optimal spacings and interconnections); the second, conductivity and magnetic measurements of our fluorinated materials show superconductivity above room temperature. The third solves the deleterious oxygen diffusion problem by replacing some oxygen atoms with fluorine and thereby assure thermal stability,[57,58] incidentally demonstrating thereby that the oxygen mobility is not connected with the superconducting mechanism. The fourth, the use of dopant amounts of fluorine to get over 90% oriented crystals as against the conventional non-fluorinated random 1,2,3 crystallites. This in principle solves the important problem of achieving high critical current densities. This has been confirmed by Mankiewich et al.[63]

V.   CONCLUSION

In conclusion, for Heinz's and my sake, I hope that the OTS can be found to be the superconductor that he suggests. Perhaps we have been "speaking prose," as Molière said, all the time without knowing it.[64] Certainly, my early transition metal oxide and chemical modification work need revisiting.

It is well known that metallic materials can go superconducting at low temperatures. I can see no reason why nonequilibrium metallic states produced by excitation, including double injection, photoexcitation and space charge accumulation, cannot do the same. Therefore, I suggest that these are interesting systems for exploring high-temperature superconductivity as well as new device structures.

In any case, there is no unique material that is accidentally found in crystal form with the right chemical configurations and spatial relationships that will give us the best high-temperature superconductors. What is needed is a basic understanding of the mechanism and the "freezing in" of metastable configurations. What I have offered here is an expansion of my approach to amorphous materials which I feel can extend the Rosetta Stone of understanding local order to deciphering high-temperature superconductivity. Its universal stereochemical principles should be useful in synthesizing new materials which help make room-temperature and above superconductors possible and practical. Valency and local coordination control are key to the future as they have been to our past. The freedom of lone pairs to adjust optimally and interactively to their local environment, the importance of separate but interactive local order, defects, controlled carrier density, the design advantages of utilizing multi-elemental synthetic materials, and the stereochemistry involved in all of the above, are not only the basis for our work in amorphous materials, but can be viewed as means for the understanding and advancement of high-temperature superconductors which after all have varied configurations and positional, translational, compositional and interfacial disorder.

Acknowledgements

I acknowledge with gratitude the collaboration through the years of David Adler. Working with him was such a pleasure. I miss him deeply. I thank Stephen Hudgens for his collaboration and contributions to the previously published high temperature superconductor model (also for discussions on the Ovonic threshold switch) and Richard Lintvedt and David Rorabacher for their contributions to it, particularly in assuring its chemical soundness. The fluorinated high-temperature superconductor experimental work owes much to Rosa Young's collaboration. My appreciation to Hellmut Fritzsche not only for his discussions, suggestions, advice and encouragement on the work discussed here but also for our collaboration over the past 25 years. My thanks to Iris for her loving support and continuous help.

Postscript

Up until the very last, I.I. Rabi was excited about physics. His encouragement and support has meant so much to me. Our last conversations were about this new high-temperature superconductivity work.

References

1. H.K. Henisch, "Rectifying Semiconductor Contacts," Clarendon Press, Oxford (1957).

2. H.K. Henisch and R.W. Pryor, "Mechanism of Ovonic Threshold Switching," Solid State Elec. 14:765 (1971).

3. R. W. Pryor and H.K. Henisch, "First Double Pulse Transient Study of the ON-State (TONC)," J. Noncryst. Solids 7:181 (1972).

4. H.K. Henisch, R.W. Pryor and G.J. Vendura, "Characteristics and Mechanisms of Threshold Switching,: J. Noncryst. Solids 8-10:415 (1972).

5. H.K. Henisch, W.R. Smith and M. Wihl, "Field-Dependent Photo-Response of Threshold Switching Systems," in: Proc. the 5th Intl. Conf. on Amorphous and Liquid Semiconductors, Garmisch-Partenkirchen, Germany, J. Stuke and W. Brenig, eds., Taylor and Francis, London, 567 (1974).

6. D. Adler, H.K. Henisch and N. Mott, "The Mechanism of Threshold Switching in Amorphous Alloys," Rev. Mod. Phys. 50:209 (1978).

7. S.R. Ovshinsky, "Reversible Electrical Switching Phenomena in Disordered Structures," Phys. Rev. Lett. 21:1450 (1968).

8. See S.R. Ovshinsky, "Fundamentals of Amorphous Materials," in: Physical Properties of Amorphous Materials, D. Adler, B.B. Schwartz and M.C. Steele, eds., Institute for Amorphous Studies Series, Plenum Publishing Corporation, New York (1985) for early references.

9. H.K. Henisch, "Threshold Switching--A Form of Superconductivity?", unpublished (1987).

10. S.R. Ovshinsky, "Principles and Applications of Amorphicity, Structural Change, and Optical Information Encoding," in: <u>Proc. 8th Intl. Conf. on Amorphous and Liquid Semiconductors</u>, Grenoble, France (1981): <u>J. de Physique</u>, Colloque C4, supplement au no. 10, 42:C4-1095 (1981).

11. S.R. Ovshinsky, "The Chemical Basis of Amorphicity: Structure and Function," <u>Revue Roumaine de Physique</u> 26:893 (1981); also in: <u>Disordered Materials: Science and Technology</u>, Selected Papers by S.R. Ovshinsky, D. Adler, ed., Amorphous Institute Press, Bloomfield Hills, MI (1982). (Grigorovici Festschrift.)

12. S.R. Ovshinsky, "The Shape of Disorder," <u>J. Noncryst. Solids</u> 32:17 (1979). (Mott Festschrift.)

13. S.R. Ovshinsky and D. Adler, "Local Structure, Bonding, and Electronic Properties of Covalent Amorphous Semiconductors," <u>Contemp. Phys.</u> 19:109 (1978).

14. S.R. Ovshinsky and H. Fritzsche, "Amorphous Semiconductors for Switching, Memory, and Imaging Applications," <u>IEEE Trans. on Electron Devices</u> ED-20:91 (1973).

15. I wish to thank Morrel Cohen for his clarifying comments on the negative correlation argument and discussion of bipolarons.

16. J. Bardeen, L.N. Cooper and J.R. Schrieffer, "Theory of Superconductivity," <u>Phys. Rev.</u> 108:1175 (1957).

17. S.R. Ovshinsky and K. Sapru, "Three-Dimensional Model of Structure and Electronic Properties of Chalcogenide Glasses," in <u>Proc. 5th Intl. Conf. on Amorphous & Liquid Semiconductors</u>, Garmisch-Partenkirchen, Germany 1973; J. Stuke and W. Brenig, eds., Taylor and Francis, London (1974).

18. I hope that Heinz forgives me my emphasis on lone pairs. Certainly other carriers can initiate and make up a highly dense plasma, but the presence of lone pairs also in nonchalcogenide materials such as in group V, albeit not as pronounced or available, still must be taken into account. In any case, we have seen some forms of threshold switching in nonchalcogenide materials early at ECD and later at Penn State (K. Homma, H.K. Henisch and S.R. Ovshinsky, J. Noncryst. Solids 35&36:1105 (1980)) which indicates to me that there is a possibility of the critical on-state plasma being present in a spectrum of materials. However, there is no question that the switching mechanism is best seen and most stable in the lone-pair chalcogenides since the excitation process occurs in the nonbonded lone pairs rather than in the structural bonds as it does in other materials. It is in these materials that the effective negative correlation energy reigns supreme and that the volumetric control of the constant current density of the filament is best expressed.

19. S.R. Ovshinsky, "Amorphous Materials As Interactive Systems," Proc. 6th Intl. Conf. on Amorphous & Liquid Semiconductors, Leningrad, 1975: Structure and Properties of Non-Crystalline Semiconductors, B.T. Kolomiets, ed., Nauka, Leningrad (1976); and oral presentation (see H. Fritzsche, Proc. 6th Intl. Conf. on Amorphous & Liquid Semiconductors, Leningrad, 1975: Electronic Phenomena in Non-Crystalline Semiconductors, B.T. Kolomiets, ed., Nauka, Leningrad (1976)).

20. S.R. Ovshinsky, "Lone-Pair Relationships and the Origin of Excited States in Amorphous Chalcogenides," Proc. of the Intl. Topical Conference on Structure and Excitation of Amorphous Solids, Williamsburg, Virginia (1976).

21. S.R. Ovshinsky, "Localized States in the Gap of Amorphous Semiconductors," Phys. Rev. Lett. 36:1469 (1976).

22. Kastner, Adler and Fritzsche took up this theme of the one- and three-electron pairs and made an elegant and important model based upon valence alternation pairs: M. Kastner, D. Adler and H. Fritzsche, "Valence-Alternation Model for Localized Gap States in Lone-Pair Semiconductors," Phys. Rev. Lett. 37:1504 (1976).

23. S.R. Ovshinsky, "The Quantum Nature of Amorphous Solids," in: Disordered Semiconductors, M.A Kastner, G.A. Thomas and S.R. Ovshinsky, eds., Institute for Amorphous Studies Series, Plenum Publishing Corporation (1987). (Fritzsche Festschrift.)

24. S.R. Ovshinsky, "Chemistry and Structure in Amorphous Materials: The Shapes of Things to Come," in Physics of Disordered Materials, D. Adler, H. Fritzsche and S.R. Ovshinsky, eds., Institute for Amorphous Studies Series, Plenum Publishing Corporation (1985). (Mott Festschrift.)

25. M.H. Cohen, H. Fritzsche and S.R. Ovshinsky, "Simple Band Model for Amorphous Semiconducting Alloys," Phys. Rev. Lett. 22:1065 (1969).

26. W.A. Little, "The Possibility of Synthesizing an Organic Superconductor," Phys. Rev. A 134:1416 (1964).

27. V.L. Ginzburg, "On Surface Superconductivity," Phys. Lett. 13:101 (1964).

28. D. Allender, J. Bray and J. Bardeen, "Model for an Exciton Mechanism of Superconductivity," Phys. Rev. B 7:1020 (1973).

29. We cannot locate the reference. Dave, how we miss your encyclopedic memory!

30. N. Sakai and H. Fritzsche, "Semiconductor-Metal and Superconducting Transitions Induced by Pressure in Amorphous As_2Te_3," Phys. Rev. B 15:973 (1977).

31. Internal ECD reports, 1975 and 1977. Samples remeasured at the Francis Bitter National Magnet Lab, Report dated July 1982–June 1983, p. 118.

32. H. Sadate-Akhavi, J.T. Chen, A.M. Kadin, J. E. Keem and S.R. Ovshinsky, "Observation of RF-Induced Voltages in Sputtered Binary Superconducting Films," Solid State Commun. 50:975 (1984).

33. A.M. Saxena, J.E. Crow and M. Strongin, "Coherent Properties of a Macroscopic Weakly Linked Superconductor," Solid State Commun. 14:799 (1974).

34. S.R. Ovshinsky, "Resistance Switches and the Like," U.S. Patent No. 3,271,719 (original filed June 21, 1961), issued September 6, 1966.

35. See S.R. Ovshinsky and I.M. Ovshinsky, "Analog Models for Information Storage and Transmission in Physiological Systems," Mat. Res. Bull. 5:681 (1970) for early references. (Mott Festschrift.)

36. S.R. Ovshinsky, "Chemical Modification of Amorphous Chalcogenides," in: Proc. of 7th Intl. Conf. on Amorphous and Liquid Semiconductors, Edinburgh, Scotland (1977).

37. R.A. Flasck, M. Izu, K. Sapru, T. Anderson, S.R. Ovshinsky and H. Fritzsche, "Optical and Electronic Properties of Modified Amorphous Materials," in Proc. 7th Intl. Conf. on Amorphous and Liquid Semiconductors, Edinburgh, Scotland (1977).

38. S.R. Ovshinsky, "The Chemistry of Glassy Materials and their Relevance to Energy Conversion," in: Proc. Intl. Conf. on Frontiers of Glass Science, Los Angeles, California; J. Noncryst. Solids 42:335 (1980).

39. B.T. Kolomiets, V.L. Averyanov, V.M. Lyubin and O.Ju. Prikhodko, "Modification of Vitreous As_2Se_3," Solar Energy Mats. 8:1 (1982). (Ovshinsky Festschrift.)

40. E.A. Davis and E. Mytilineou, "Chemical Modification of Amorphous Arsenic," <u>Solar Energy Mats</u>. 8:341 (1982). (Ovshinsky Festschrift.)

41. Hamakawa called chemical modification "sensational." H. Okamoto and Y. Hamakawa, "Gap States in Amorphous Semiconductors," <u>J. Noncryst. Solids</u> 33:225 (1979).

42. S.R. Ovshinsky, "Polymeric Semiconductors," Lecture Notes For "Recent Advances in Polymeric Materials (March 1977).

43. In the early 1960's, we called the OTS the Quantrol. J.R. Bosnell, "Amorphous Semiconducting Films," in Active and Passive Thin Film Devices, T.J. Coutts, ed., Academic Press (1978).

44. How orbitals interact differently, perhaps fractally, in amorphous materials is discussed, for example, here and in reference 22. S.R. Ovshinsky, "Basic Anticrystalline Chemical Configurations and Their Structural and Physical Implications," <u>J. Non-Cryst. Solids</u> 75:161 (1985).

45. S.R. Ovshinsky, S.J. Hudgens, R.L. Lindvedt and D.B. Rorabacher, "A Structural Chemical Model for High T_c Ceramic Superconductors," <u>Modern Physics Letters B</u>, Vol. 1, Issue 7/8 (October/November 1987).

46. (a) J.G. Bednorz and K.A. Muller, "Possible High T_c Superconductivity in the Ba-La-Cu-O System," <u>Z. Phys. B - Condensed Matter</u> 64:189 (1986); (b) M.K. Wu, J.R. Ashburn, C.J. Tong, P.H. Hor, R.L. Wong, L. Gao, Z.J. Huang, Y.Q. Wang and C.W. Chu, "Superconductivity at 93K in a New Mixed-Phase Y-Ba-Cu-O Compound System at Ambient Pressure," <u>Phys. Rev. Lett</u>. 58:908 (1987) and P.H. Hor, L. Gao, R.L. Meng, Z.J. Huang, Y.O. Wang, K. Forster, J. Vassiliow and C.W. Chu, "High-Pressure Study of the New Y-Ba-Cu-O Superconducting Compound System," <u>Phys. Rev. Lett</u>. 58:911 (1987); (c) News reports on bismuth materials.

47. A.R. Von Hippel, "Molecular Science and Molecular Engineering," The Technology Press of M.I.T and John Wiley & Sons, Inc., New York (1959), p. 259.

48. H.A. Jahn and E. Teller, "Stability of Polyatomic Molecules in Degenerate Electron States," Proc. Roy. Soc. A161:220 (1937).

49. A.I. Braginski, "Carrier Density Measurement Using Hall Effect," in: Proc. Intl. Workshop on Novel Mechanisms of Superconductivity, V. Kresin and S.A. Wolf, eds., Plenum Press, New York (1987).

50. W.W. Warren, Jr., R.E. Walstedt, G.F. Brennert, G.P. Espinosa and J.P. Remeika, "Evidence for Two Pairing Energies from Nuclear Spin-Lattice Relaxation in Superconducting $Ba_2YCu_3O_{7-\delta}$," Phys. Rev. Lett. 59:1860 (1987).

51. S.R. Ovshinsky, R.T. Young, D.D. Allred, G. DeMaggio and G.A. Van der Leeden, "Superconductivity at 155K," Phys. Rev. Lett. 58:2579 (1987).

52. X.R. Meng, Y.R. Ren, M.Z. Lin, Q.Y. Tu, Z.J. Lin, L.H. Sang, W.Q. Ding, M.H. Fu, Q.Y. Meng, C.J. Li, X.H. Li, G.L, Qiu and M.Y. Chen, "Zero Resistance at 148.5K in Fluorine Implanted Y-Ba-Cu-O Compound," Solid State Commun. 64:325 (1987).

53. Z.X. Zhao, Academia Sinica, Beijing, personal communication.

54. J.H. Kung, in: Proc. 1987 Symposium on Low-Temperature Physics, September 7-8, 1987, Hsin-Chu, Taiwan.

55. P.T. Wu, R.S. Liu, S.M. Suhng, Y.C. Chen and J.H. Kung, "Possibility of High T_c Copper Fluoride Oxide Superconductors," presented at the 1987 Materials Research Society Meeting, November 30-December 5, 1987, Boston, MA.

56. C. Krontiras, personal communication.

57. S.R. Ovshinsky, R.T. Young, B.S. Chao, G. Fournier and D.A. Pawlik, "Superconductivity in Fluorinated Copper Oxide Ceramics," presented at the Intl. Conf. on High-Temperature Superconductivity, July 29-30, 1987, Drexel University, Philadelphia, PA; in: <u>Proc. of the Drexel Intl. Conf. on High-Temperature Superconductivity</u>, S. Bose and S. Tyagi, eds., World Scientific Publishing Co., Singapore (January 1988).

58. S.R. Ovshinsky, "Superconductivity at 155K and Room Temperature," presented at Superconductors in Electronics Commercialization Workshop, San Francisco, California, September, 1987.

59. R.T. Young, S.R. Ovshinsky, B.S. Chao, G. Fournier and D.A. Pawlik, "Superconductivity in the Fluorinated YBaCuO," presented at the Materials Research Society Meeting, November 30-December 5, 1987, Boston, MA.

60. We have one 370K measurement.

61. V.L. Ginzburg, "High-Temperature Superconductivity: Some Remarks," November 1987, to be published in <u>Progress in Low-Temperature Physics</u>.

62. S.R. Ovshinsky, in Collection of papers on amorphous materials in honor of Professor David Adler, China; to be published.

63. P.M. Mankiewich, J.H. Scofield, W.J. Skocpol, R.E. Howard, A.H. Dayem and E. Good, "Reproducible Technique for Fabrication of Thin Films of High Transition Temperature Superconductors," <u>Appl. Phys. Lett</u>. 51:1753 (1987).

64. Moliere, Le Bourgeois Gentilhomme (1670).

65. Happy 65th Birthday, Heinz!

PHENOMENOLOGY OF ANTIAMORPHOUS ORDER

A. H. Madjid, W. F. Anderson, Jr. and R. L. Osgood

Department of Physics
The Pennsylvania State University
University Park, PA 16802

and

T. Madjid

Department of the Air Force
Headquarters Electronic Systems Division
Hanscom Air Force Base, MA 01731

INTRODUCTION

When dealing with the phenomenology associated with aggregates of matter, it is useful to appeal to a taxonomy that is based on concepts of order.

Starting with ideal gases, in which both near, and far orders are identically absent, the classification may be carried through to Bravais lattices which may be regarded as opposite extremes since in these aggregates these orders are identically perfect.

But more complex structures, including some that are currently under intensive investigation, may also be fitted into the homology.

Thus, by allowing order identity to be replaced by order difference, a number of complex, but homologous aggregates may be invoked. In crystalline layer structures, for instances, the near orders are determined by the respective repetitivities of the atomic arrays that constitute the different layers, whereas far order will be given by the identity period of the lattice, that is, the dimension of one layer pair, (or unit group in more elaborate structures).

For amorphous structures the present classification will reflect conventional wisdom by characterizing the state as possessing near, but not far order.

An interesting question arises at this point: does a conjugate to the amorphous state exist, and if such a state indeed exists, will it exhibit a significant phenomenology? Homology considerations mandate that such structures be distinguished by the absence of near, and the presence of far order and, as it turns out, it is experimentally quite simple to produce such structures.

The present paper is a summary of the efforts in the Thermionic Emission Laboratory to produce and study aggregates which, by analogy, may be called antiamorphous structures.

The experimental purpose was:

(1) to construct well defined multilayer arrays of two different substances a and b in superlattice sequence (...ababab...), (2) to deposit the atoms (or molecules) in each individual layer in sites that are distributed as randomly as possible, and (3) to make the layers as similar as possible to every other layer of its own kind. (1) and (3) yield a far order with identity period ab, and (2) assures the lack of intralayer near order.

APPARATUS

The multilayer stacks were produced by evaporating alternate layers of silver (gold and aluminum were also used), and silicon monoxide onto flat or cylindrical substrates (Aℓ, Cu, glass, quartz).

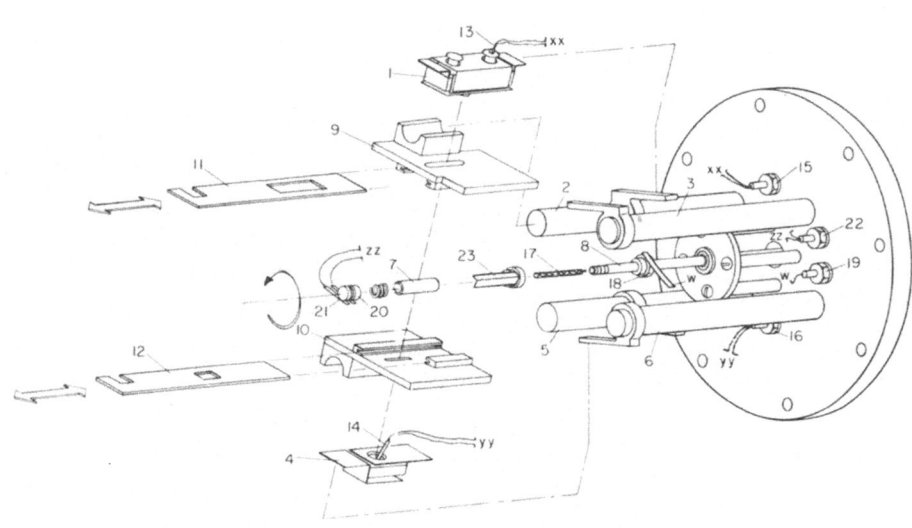

(1)	SiO gun	(13)	SiO gun thermocouple
(2)	SiO gun current	(14)	Ag gun thermocouple
(3)	SiO gun current	(15)	SiO gun thermocouple
(4)	Ag gun	(16)	Ag gun thermocouple
(5)	Ag gun current	(17)	Substrate heater
(6)	Ag gun current	(18)	Substrate heater bushing
(7)	Substrate	(19)	Substrate heater current
(8)	Substrate holder	(20)	Thermistor nut assembly
(9)	SiO mask	(21)	Thermistor bushings
(10)	Ag mask	(22)	Thermistor feedthrough
(11)	SiO gun shutter	(23)	Substrate mask
(12)	Ag gun shutter		

Fig. 1. Apparatus for producing silver-silicon monoxide superlattice stacks.

The apparatus used for the purpose is described in Fig. 1 which shows the silver and the silicon monoxide evaporator guns, the shutter systems for defining and interrupting the two vapor beams, and the rotary substrate assembly. (Shown is a cylindrical substrate, the flat substrates were mounted on cylindrical substrate holders).

NATURE OF SAMPLES

The geometric structure of the samples used in transport measurements are shown in Fig. 2. Those used in optical measurements were deposited on flat quartz substrates omitting the inner and outer electrical contact layers.

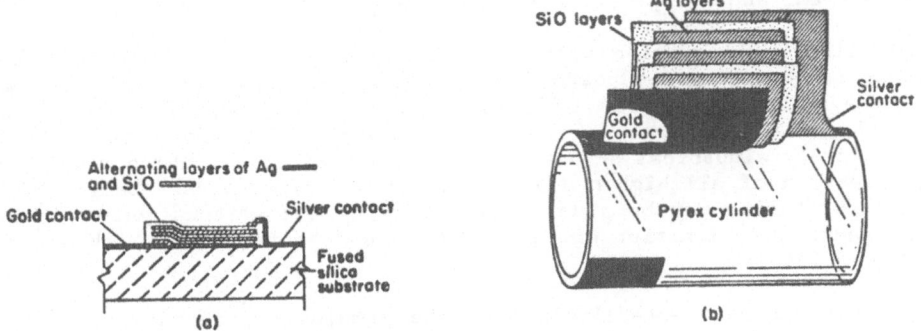

Fig. 2 Geometrical structure of the superlattice stacks.

Three independent methods were used to characterize the superlattice stacks and determine layer thicknesses and identity periods. The weighing method depended on an evaporation rate calibration of the two evaporation guns. The layer thicknesses and the identity period was, then, calculated from these calibrations, the evaporation time, and the total number of revolutions.

The use of the multiple—beam interference method developed by Tolansky (1948, 1960, 1969) permitted more precise measurement of the total stack thicknesses. In this method a measurement substrate was mounted on the substrate holder. On one part of this substrate no deposit is made, the total stack is deposited, on the next part of the wafer, and on a third part only the silicon monoxide is evaporated by blocking out the silver beam. The dimensions of the steps is then determined by coating all parts with a reflecting layer of silver in a separate evaporation cycle, and by subsequently placing a partially silvered glass flat over the one, and then over the other step. The dimension of the stack, the layer thicknesses and the identity period are, then, computed from the observed Fizeau fringe system shifts under yellow sodium doublet illumination and from the revolution count.

But the above measurements, even if the evaporation rate and the rotation is held constant, do not unequivocally prove that the stacks are characterized by the computed identity period. It is, for instance, conceivable that diffusion during and after evaporation may effectively obliterate the desired regularity. The only direct identity period test is to subject the stacks to a diffraction experiment involving an incident wave field. This was accomplished by using a Siemens Kratky small angle x—ray camera (Kratky, 1954) in conjunction with filtered

Table I   Comparison of the superlattice constant determined by
               Tolansky's method and x-ray diffraction.

| Sample # | Superlattice Constant | | Difference |
	Tolansky's Method	x-ray Diffraction	
80	47Å	53Å	11%
81	114Å	97Å	15%
85	197Å	190Å	4%
84	346Å	340Å	2%

Cu K_α radiation.   The results for a few samples are shown in Table I where the values for the identity period obtained by the last two methods are compared, and a typical first order peak is shown in Fig. 3.

A notable observation within the context of these measurements was absence of second order peaks (except in a very few cases).   This may indicate that interlayer diffusion is an important factor.   Such diffusion can lead to an electron density distribution that is approximately sinusoidal across the succession of layers which results in a suppression of all higher order peaks beyond first order (Ball, 1971). Indeed, diffusion may be studied by observing the attenuation of higher order peaks as a function of time (DuMont and Youtz, 1984; Dinklage and Frerichs, 1963).

This experiment establishes that the samples are characterized by a diffraction measurable identity period, that is, by a long range order within the initial definition.   There remains to discuss the nature of the intralayer structure.

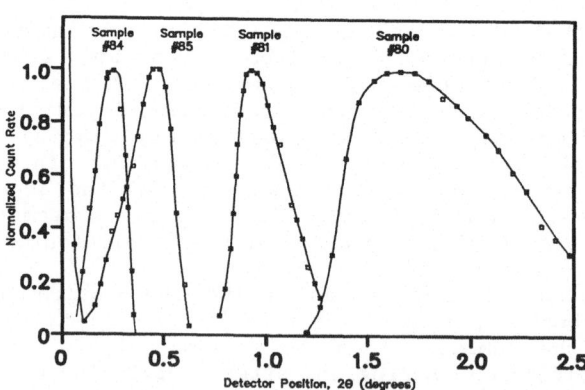

Fig. 3.   First order x-ray diffraction peak for four samples.

An extensive literature exists on the properties of extremely thin silver films (e.g., Boiko, Synelnikov and Kopach, 1971; Doremus, 1966; Hartman, 1963; Sennett and Scott, 1985), and also vapor deposited silicon monoxide (Allam and Pitt, 1967; Howson and Taylor, 1971; White and Roy, 1964). Our own preliminary electron microscopic studies tended to confirm most of the findings and conclusions of the investigators cited.

Thick layers of silver evaporated in a good vacuum are aggregates of a randomly distributed cluster type. Such structures contain voids and occlusions, but numerous bridges will exist to yield many continuous paths across the solid and the transport and optical properties will not, as a rule, exhibit pronounced anomalities. This tends to be true even for layer thicknesses below about 100 Å provided the microstructure of such films is taken into account. Such layers have an island and land structure and may be regarded as Maxwell—Garnett aggregates (Maxwell—Garnett, 1904, 1906; Cohen et al., 1973). Thus, small particle effects will have to be allowed for when adopting various transport and optical parameters.

There are three main effects to be considered when dealing with layers less than 100 Å thick. To begin with, the effects of the microstructure, which is an aggregate of spheroidal particles of an average size of about 20 Å, is accounted for in the generalized Maxwell—Garnett theory (Cohen et al., 1973). The second effect pertains to changes in solid state properties (e.g., changes in the chemical potential, scattering characteristics etc.), of small regions ($\Delta V \lesssim 100$ Å), when such regions are not surrounded by the parent bulk material but by a substance of different composition. The third phenomenon relates to those changes brought about by isolating small volumes ΔV. This leads to size quantization if ΔV becomes sufficiently small and to changes in scattering characteristics when the mean free path begins to exceed the linear dimensions ΔV. Yet, although the effects on the pertinent transport and optical parameters are by no means negligible below about 100 Å, the character of the layers remains essentially that of metallic silver down to thicknesses of about 10 Å when more radical changes in electrical conductivity and optical constants begin to occur.

For the silicon monoxide layer the situation will be similar as far as deposition geometry is concerned. As a rule SiO deposits as silicon spheroidal particles of about 20 Å on the average in a matrix of silicon dioxide. Most remarks made in the previous paragraph will also apply to this case except that, since silicon is a semiconductor and not a metal, the layer will reflect semiconducting and not metallic character.

Our view, based on the present evidence, is that the vapor deposited stacks have the following structure. The silicon monoxide layers are probably composed of an aggregate of about 50% Si:50% SiO$_2$ in which 10—50 Å diameter silicon clusters are interspersed with silicon dioxide islets of similar dimension. Following this layer there should be an interfacial layer in which the above composition is augmented by the addition of globules of silver. The composition of the interfacial layer is probably not uniform. Starting at the SiO layer side with roughly 50% Si:50% SiO; changing to an, as yet, undeterminate mixture of Si, SiO$_2$, and Ag in the middle; and tapering off to 100% Ag toward the silver layer side. The silver layer is probably composed of coalescent lands of silver consisting of spheres in the above size range. Following the silver layer is another interfacial layer, with reverse order of composition to the one described thus completing the identity period.

To summarize, it is certain that there will exist at least moderate degree of randomness as far as the intralayer regions are concerned and that the identity period will, therefore, partially derive from the interlayer order that results when these disordered two kinds of layers are stacked in an orderly array and the conclusion, therefore, is that the experimental stack produced tends to meet the criteria that distinguishes the antiamorphous state.

NATURE OF THE ANTIAMORPHOUS STATE

Within the realm of the first principle description of the nature of solids it may be shown, by making use of Floquet's or Bloch's theorem, that lattice periodicity is a sufficient and necessary condition, within the one electron approach, for the electronic states within the solid to be describable by a complete set of energy eigenfunctions, the Bloch functions,

$$\psi_{n,\vec{k}}(\vec{r}) = \mu_{n,\vec{k}}(\vec{r})e^{i\vec{k}\cdot\vec{r}} \quad , \tag{1}$$

These functions are plane running wave solutions that are modulated by the continuous lattice–periodic function $\mu_{n,k}(\vec{r})$. The dispersion of the energy eigenvalues in k–space is such that the states aggregate in quasi–continuous bands 1, 2, ...n ... which are separated by forbidden energy regions. The dispersion $E_n(\vec{k})$, or the band structure of the solid in question, is determined by the nature of the atomic wavefunctions of the atoms that compose the lattice of the solid in question and also by the periodicity and the symmetry of the lattice itself. Clearly, the key to all solid state phenomena that rest in the behavior of electrons in crystalline solids will be based upon the two above entities. But since most other material properties, such as for example lattice vibrations, depend upon the behavior of various wave fields in the periodic environment of the crystalline structure of the solid, the same conclusion, that the key to the properties of solids ultimately rests in the periodicity of the lattice and the nature of the atoms (or molecules) that compose the solid, will apply with almost universal validity to the crystalline condensed state.

But, for the case of antiamorphous superlattice several additional points need to be considered. A salient point is that the electronic potential across the superlattice can not be strictly periodic since the intralayer structure is random. This will conflict with the requirements of Floquet's and, therefore, with Bloch's theorem.

However, if the number of atoms in each layer type, n_a and n_b, is sufficiently large, the law of large numbers will assure that the average atom separation is identical for each layer a and each layer b, and it may well be legitimate to apply Bloch's theorem when dealing with space cooperative bulk phenomenons such as, for instance, appropriate transport and optical measurements that involve summing contribution from an appreciable part of the bulk. Vapor deposited aggregates have, after all, sample–to–sample reproducible physical properties. Such solids may be thought of as a crystalline order perturbed by a positional randomness of comparatively small standard deviation which is frozen in by the evaporation process since kT over the binding energy may serve as a rough measure of this deviation (with T chosen nearer to the melting point than to room temperature). The positional randomness expected will probably be comparable to that exhibited by crystalline order at the specific temperature when viewed at a given moment of time.

We will, therefore, make the tacid assumption that as for the case of thermal vibrations, where a time average is invoked to re—establish periodicity, it is legitimate to use the positional space average to re-establish periodicity in antiamorphous structures in order to interpret electronic itinerancy and allied bulk phenomena in such aggregates.

The consequences of such a view would, of course, be that the bulk properties of two ingredients a and b would tend to change drastically by layering them in superlattice order. To test whether this is indeed true for the present experiments, three samples were prepared with similar structures except that in the first sample only Ag was consecutively layered, in the second sample only SiO was layered, and only in the third sample both Ag and SiO were layered in superlattice order. The resistivity—temperature plot of the first sample showed typical metallic conductivity with a room temperature resistivity in the 10^{-6} Ωcm range and a positive temperature coefficient of about 0.002 per degree. The second sample showed a typical semiconducting conductivity characterized by a linear logarithm of conductivity versus the inverse temperature plot yielding an activation energy of 0.77 eV and a room temperature resistivity of about 3×10^{14} Ωcm. The conductivity plot of the third superlatticed sample also showed semiconducting character but with activation energy of only 0.097 eV and a room temperature resistivity of 5×10^6 Ωcm. (Contact resistance measurements were also made to make sure that the resistances measured did not reside in the contacts.) Since the resistance of the silver layer is negligible compared to that of the silicon monoxide the expectation would be, assuming that layering is a linear effect, that the superlattice stack would virtually have the same resistance as the silicon monoxide sample because in the former stack the layer resistances are in series and may under the assumption simply be summed. But, this is evidently not the case. Indeed the process of layering has the effect of lowering the resistance of the component materials by eight order of magnitude, and of lowering the activation energy by a factor of eight. Such decreases may, indeed, be brought about by layering if the bandstructure of the superlattice is determined in significant part by the latter's relativity large identity period.

In the next section the relation between superlattice identity period and its bandstructure is examined and illustrated by using a simple Kronig—Penney model. This discussion is general in the sense that it applies to all one dimensional superlattices and not to antiamorphous layer aggregates exclusively.

BAND STRUCTURE MODULATION IN SUPERLATTICES

Layered superlattices, of necessity, incorporate interfaces at each layer boundary. But, the contact between two solids is a region in which the occurrence of nonitinerant trapping states may, indeed, be expected with great probability. To begin with, the interface is characterized not only by a unidirectional, but by a bidirectional interruption of material identity; and this can give rise to various interfacial surface states. But the contact, as the transition region between two mismatched solid orders, is generally marked by disorder, which makes the occurrence of local defect states also probable. Boundaries, finally, are usually impurity rich regions because foreign species are often trapped in the interface by occlusion, and preferentially diffuse into the contact during manufacturing. This, of course, may also cause localization impurity states to appear in this region. This complexity does not make for any easy theoretical description, but it does make the appearance of trapping states in interfaces the rule, rather than the exception. This is confirmed in the rather voluminous literature that deals with slow and fast states associated with contacts between metals, insulators, semiconductors, empty space, and gaseous ambiences in various combinations.

But, to turn to superlattices (that is arrays in which the repetitivity of n_a atoms of substance a, followed by the repetitivity of n_b atoms of substance b, are in turn, ordered into a periodic layer structure; or arrays such as molecular crystals in which the molecular basis itself has chainlike structure), such doubly periodic arrays must, of necessity, contain a periodic succession of interfaces each capable of trapping charge carriers in local energy states. The occupation of such donor and acceptor states will be governed by the location of the Fermi level within the structure. An unbalanced charge occupying these states will, however, not be random, but in periodic interfacial sites, and this will give rise to an extraneous periodic potential which will superimpose on that given by the identity period structure of the superlattice. The maximum field strength at the boundary $F_B{}^{max}$, due to such an occupation may be estimated roughly from,

$$F_B = \frac{e \eta_B}{\varepsilon \varepsilon_0} \qquad (2)$$

with e the electronic charge, ε the effective dielectric constant of the respective layer, ε_0 the permitivity of free space, and η_B the number of unbalanced charges in boundary states per unit boundary area. Assuming $\varepsilon = 4$, and an upper limit for η_B of about one half the number of boundary atoms per unit area, or $\eta_B \approx 10^{14}$ cm^{-2}, the estimate for $F_B{}^{max}$, thus, turns out to be a large 10^8 volt/cm. The difference in the electronic potential energy in the layer bulk and at the boundary, $e(\Delta V_{BL})$, will be from a fraction of an electron volt for metallic layers, to an upper limit, that is roughly equal to the band gap value, ΔE_G, for semiconducting and insulating layers. The maximum $e(\Delta V_{BL})$ to be expected may, thus, be of the order of several eV. The thickness of the periodic space charge regions, finally, in which most of the change in electronic potential between layer bulk and boundary occurs, can vary from less than an atom distance to several hundred atom distances for metals and insulating materials respectively. These are, evidently, manifestations that are not trivial. The exchange of charge between boundaries and bulk, after all, leads to manifold phenomena in various p—n junction, tunnel, and M—O—S devices and there is no reason to expect that this effect will be less pronounced for periodic arrays of interfacial charge.

The salient point of the present argument, however, is that in contra-distinction to such nonrepetitive boundary layer structures, or to ordinary insulators—where trapping states are distributed randomly throughout the crystal bulk and where the accumulation of charge in these trapping states may be treated as a charge continuum in the derivation of the well known ordinary insulator space charge effects—trapping states in superlattices will be distributed periodically throughout the bulk. Any change in trap occupancy will, therefore, result in a periodic modulation of the electronic potential through the superlattice and this, in turn, if sufficiently pronounced, will lead to a modification of the band structure, or the dispersion $E(\vec{k})$, of the assembly. Because of this, nonequilibrium charge accumulation (or depletion) in the interfacial traps will alter the electronic transition spectrum; it will also alter the free energy of the assembly. The entire superlattice will, thus, tend to act as a giant molecule which may be lifted from some equilibrium ground state to excited states by any agency that can alter the occupancy of the periodic trapping sites along the structure and in which it will persist until the excess charge leaks out of the traps. Illumination, changes in temperature, application of strong electric fields, and charge carriers transport through the structure may all alter the dispersion spectrum and this modulation of the band structure may indeed include, in principle, transitions from and to metallic, insulating, and semiconducting states.

The following model has been chosen to illustrate some of these contentions. A succession of δ-function representing potential wells P_1 and P_3, each of which may be thought of as having been arrived at by averaging over the potential structure of the two layers is arrayed into an iterative structure with identity period I. This basic lattice can now be perturbed by a secondary lattice of δ-functions P_2 of variable strength representing the potential iteration due to the variable accumulation of charge in local interfacial or else, intralayer states. The complete model is thus the superposition of the two structures analytically represented by

$$eV(X) = \sum_{n=\infty}^{\infty} \sum_{i=i}^{3} P_i \delta [x - (n+\beta_i)I] \quad . \tag{3}$$

P_i is the strength of each respective δ-function, and β_i marks its location within the identity period I. It is convenient to define a dimensionless strength of the delta functions as

$$\pi_i = \frac{m_e I}{\hbar^2} P_i \tag{4}$$

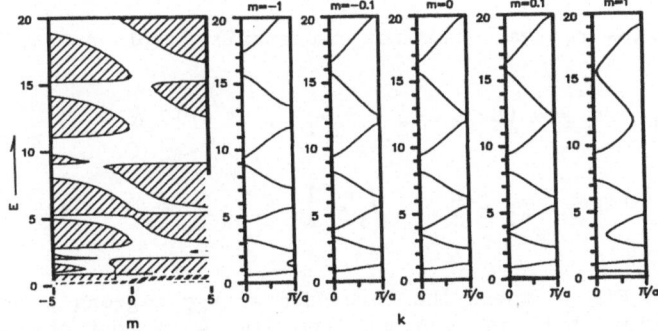

Fig. 4. The dispersion relation E(k), arising from equations (6) and (8), as a function of the intralayer charge parameter m is sketched on the right, and the resulting changes in the allowed and forbidden energy regions are shown on the left.

with m_e equal to the electronic mass. Also, to write the energy E in terms of

$$K = \left[\frac{2m_e}{\hbar^2} E\right]^{1/2} \quad . \tag{5}$$

The dispersion relation $E(k)$ is then given implicitly by, (Santana and Rosato, 1973),

$$k = \frac{1}{I} \cos^{-1} \left\{ \cos(KI) - \frac{\pi_1 + \pi_2 + \pi_3}{KI} \sin(KI) \right.$$

$$+ \frac{2\pi_1\pi_2}{(KI)^2} \sin(KI\beta_2) \cdot \sin[KI(1-\beta_2)]$$

$$+ \frac{2\pi_1\pi_2}{(KI)^2} \sin(KI\beta_3) \cdot \sin[KI(1-\beta_4)]$$

$$+ \frac{2\pi_2\pi_3}{(KI)^2} \sin[KI(\beta_3-\beta_2)] \cdot \sin[KI(1-\beta_3+\beta_2)]$$

$$\left. - \frac{4\pi_1\pi_2\pi_3}{(KI)^3} \sin(KI\beta_2) \cdot \sin[KI(\beta_3-\beta_2)] \cdot \sin(KI\beta_3) \right\} \tag{6}$$

The dispersion $E(k)$ and the spectrum of allowed and forbidden energies computed from this relation as a function of the relative periodic interfacial charge variable,

$$m_2 = \frac{\pi_2}{\pi_1} \quad , \tag{7}$$

for the illustrative, (but not necessarily representative), numerical parameters,

$$\pi_1 = \pi_3 = 1.00 \quad ; \quad \pi_2 = m_2\pi_1 \quad ; \quad I = 12 \text{ Å}$$

$$\beta_1 = 0 \quad ; \quad \beta_2 = \frac{1}{3} \quad ; \quad \beta_3 = \frac{2}{3} \quad , \tag{8}$$

is shown in Fig. 4 where the forbidden energy regions have been shaded. The conclusion that can be drawn from Fig. 4 is that the band structure for this model, indeed strongly depends on the charge in the interfaces since forbidden regions clearly appear and disappear as this charge is varied. The consequence is that transitions from and to metallic, semiconducting and even insulating states may occur as a consequence of charge accumulations in periodic local states in superlattice structures.

188

PHENOMENLOGY OF THE ANTIAMORPHOUS STATE

1. Electrical Transport and Switching

Layer structures which were used in electrical conductivity measurements, and which were found to exhibit switching characteristics had identity periods from about 25 to 170 Å. The individual layer thicknesses ranged from 20 to 90 Å for the SiO layers and from 3 to 85 Å for the Ag layers. The total number of deposited layer pairs in the samples varied from 60 to 360. Virtually all samples tested exhibited a low field conductivity characteristic of the form,

$$\sigma = \sigma_0 e^{\frac{\Delta \dot{\varepsilon}^*}{2kT}} \quad , \tag{9}$$

for applied potentials up to about 30 mV. σ is here the conductivity; σ_0 a constant pre—exponential term; and $\Delta \varepsilon^*$ the effective activation energy. The application of excessive potentials tended to alter the characteristic irreversibly.

Both $\Delta \varepsilon^*$ and σ_0 depended on the identity period and we identified this dependency within the interval $30 \leq \Pi \leq 400$ Å tentatively as,

$$\Delta \varepsilon^*(\Pi) = 1.5[1 - \exp(-2.9 \times 10^{-3}\Pi]eV \tag{10}$$

$$\sigma_0(\Pi) = 1.2 \times 10^{-10}\exp(2.9 \times 10^{-2})(Ohm\ cm)^{-1} \tag{11}$$

with Π given in Å.

For applied potentials above 30 mV and below about 1 volt the current through the samples depended on the voltage V as,

$$I = const.\ V^m \tag{12}$$

$$(m-1) = \frac{68}{\Pi} \quad , \tag{13}$$

with Π again given in Å.

Above 1 volt generally, but sometimes already below this voltage, switching phenomena would occur in many, but not all the samples. Both reversible and irreversible switching events were observed. After irreversible switching, the sample could not be returned to its initial low field conductivity state. Reversible switching occurred either slowly or rapidly. A typical slow mode switching characteristic is shown in Fig. 5a. Raising the potential across the sample rapidly, yielded the trace OA on the characteristic. Stopping the voltage sweep at point A caused a time dependent increase of the current to its ultimate value at B. This increase could be expressed as,

$$R(t) - R_B(\infty) = [R_A(0) - R_B(\infty)]\exp(-\lambda t) \quad , \tag{14}$$

with t the time, λ the time constant, and the R's defined by the respective (V/I)'s. Path AO was retraced if the voltage was swept downward immediately, (t<<λ). But BO was the trace if the voltage decrease was started only when equilibrium was established after several λ's. A similar situation developed when O was reached along BO. Recycling immediately would give OB but the equilibrium trace would be OA. Any desired trace between the two extremes could be chosen, at will, by varying the "resting time" at A along OA or at O along BO. The time constants for different samples ranged from a fraction, to many minutes. λ for the characteristic shown in Fig 5a was 1 minute.

In other samples, generally at higher voltages, the transition for A to B occurred rapidly, (λ < 50 μsec). Such samples did not have intermediate traces, but would recycle along either OB or OA depending on whether the "rest time" at O did, or did not exceed some definite value.

Some layer structures would switch to higher resistance values as shown in Fig. 5b. Others would combine the two last mentioned modes into a sequential pattern as shown in Fig. 6a. But the most peculiar and, at once, the potentially most useful switching effect observed is the multi-state pattern shown in Fig. 6b. The layer structure acts here as a multi-throw switch.

Superlattices may, with advantage, be considered as a distinct form of the condensed state. An equivalent, but separate accumulation of Ag and SiO bulk, will certainly not act in the manner described. The two components must be combined in a layer structure in order to yield the observed results. Layer structures are supercrystals in which a layer pair takes the place of the crystalline basis. If there happens to be an appreciable overlap of the Ag—state with the conduction band states of the SiO across the boundaries, then running wave solutions could exist which will extend across the entire layer structure, and which may give

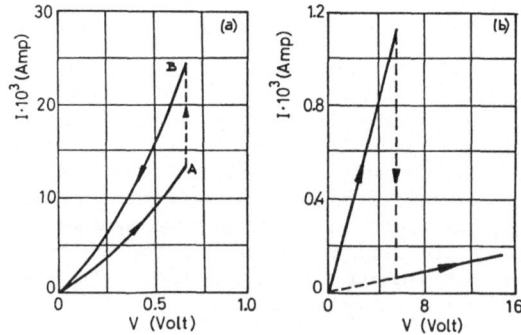

Fig. 5. Typical slow mode switching characteristic is shown in (a) Switching to higher resistance values is shown in (b).

rise to charge carrier itinerancy. The minizone dispersion, E(k), will, in part, be given by the identity period Π. Thus, Eq. (9) probably represents the activation of charge carriers from nonconducting to itinerant states. That the minizone scheme depends on Π is reflected by Eq. (10). The exponential dependence of $\sigma_0(\Pi)$ on Π may arise from the fact that samples with larger Π will have a smaller fraction of their bulk occupied by disordered boundary layers and that scattering, as a consequence, becomes less serious as Π increases. The I—V characteristics at intermediate applied potentials show that space charge effects play an important role in the charge carrier transport. This is to be expected because layer structures are in essence a microscopic version of a Maxwell—Wagner capacitor (Wagner, 1924). Because of this it is likely that space charges will predominantly accumulate in the interfacial layers. This means that a voltage dependent barrier sequence will appear in the structure at each interface. Thus it is reasonable to expect that transport may have a pronounced effect on the mini band structure of the structure and this may be the explanation of some of the switching phenomena. Equation (14) is reminiscent of the time dependence of the charge accumulation in a Maxwell—Wagner capacitor and the slowly occurring transition from A to B in the slow switching characteristic of Fig. 5b may well arise from band structure changes which are locked in step with the interfacial charge accumulation. The peculiar switching effects observed at still higher applied potentials may be due to a Wannier—Stark ladder splitting and the tunnelling of electrons from ground states in one identity cell to excited states in the neighboring cell as predicted by Kazarinov and Suris (1972). Irreversible switching, finally, may be attributed to local heating and a consequent short circuiting of the insulating layers by silver bridges. This contention is supported by the fact that $\Delta\varepsilon^*$ decreases radically after a sample switches irreversibly. Such a collapse of $\Delta\varepsilon^*$ was never observed in the reversible case even after several 100 switching cycles.

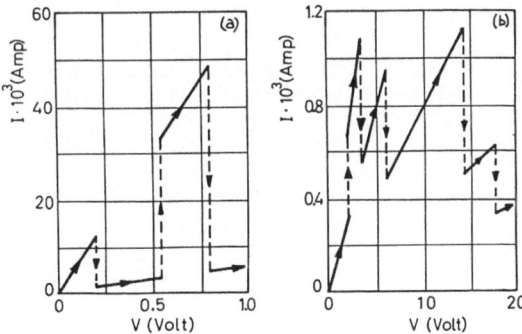

Fig. 6.   An example of a sequential switching mode is shown in (a). A multi resistance state switching mode is shown in (b).

2. Thermally Induced Switching

To investigate the effect of temperature on the electrical transport, the dc resistance of silver—silicon monoxide layer structures has been measured as a function of temperature in the range from 77 to 300 K. During this investigation it was discovered that the electrical resistance abruptly changes from the room temperature value R^{300} to a lower value $R^{\downarrow}$ at some threshold temperature $T^{\downarrow}$. When the temperature was increased again, the sample switched at a higher threshold temperature $T^{\uparrow}$ to a resistance value that is $R^{\uparrow}$ larger than R^{300}. $R^{\uparrow}$ was found not to be stable; it decayed to the room temperature value R^{300} with a time constant of the order of several hours.

The identity period of the samples investigated ranged from about 20 to 300 Å. Not all of the samples tried exhibited this switching phenomenon. The precise sample characteristics necessary for switching could not, as yet, be ascertained.

Each sample that switched could be repeatedly and reproducibly cycled many times. However, with prolonged cycling an aging phenomenon became evident which had the effect of causing both $T^{\uparrow}$ and $T^{\downarrow}$ to increase and which made $(R^{300} - R^{\downarrow})$ gradually tend to zero.

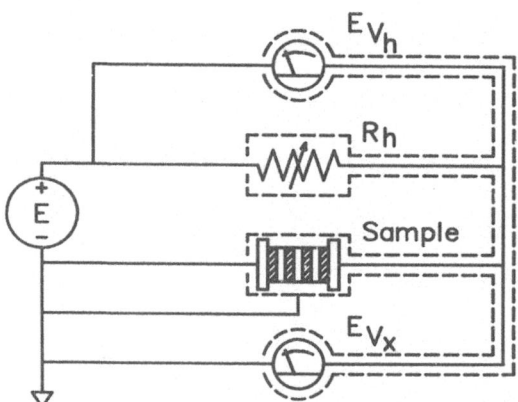

Fig. 7.   Circuit for measuring d. c. conductivity and temperature induced switching.

Such phenomena have, as far as we know, not been reported in the literature, nor do existing theories dealing with electrical transport switching phenomena such as threshold and memory switching seem to be relevant to the new thermal resistance switching effect that is described.

Many different conductivity—state switching phenomena in thin films and amorphous materials have been observed and reported in the literature and threshold switching is probably one of the most widely studied switching processes. This type of switching can be described as a switching from low conductivity state to a high conductivity state at a specific threshold voltage. The high conductivity state is maintained by a holding voltage or a holding current.

The type of switching discussed in the present work indicates that the conductivity increases as temperature is decreased in the switching region. Most of the types of switching published to date, except for the switching in TTF—TCNQ, show little dependence of the conductivity on temperature, or they show an increase in the conductivity as the temperature is increased.

The samples used in this investigation were, in general, thinner than the samples used in the switching experiments reported in literature. Another important difference is that contrary to samples used in published results, the present samples were multiply layered.

Finally, the switching in the present study showed a hysteresis in the resistance versus temperature curve. That is, the threshold temperature going from the initial high resistance state to the low resistance state was different from the threshold temperature going from the low resistance state to the high resistance state as the temperature was increased. This behavior, too, is not typical of published switching results.

The electrical circuit for measuring the conductivity (and the switching) of the superlattice samples is shown in Fig. 7. Electrometer (or potentiometer) E_{Vx} is connected directly across the sample; it measures the voltage drop across the sample. Electrometer (or potentiometer) E_{Vh} which measures the current through the sample is connected across a continuously variable resistance R_h. The emf for the circuit, E, was supplied by a potentiometer producing a continuously variable potential from one mirovolt to 1.6 volts using two Edison cells as the working battery.

The high impedance part of the circuit, R_h and the sample are carefully shielded. The low side of the sample is grounded. The possibility of varying both the applied voltage, E, and the resistance R_h makes it possible to measure the current—voltage characteristics of a great variety of samples which may not only appreciably differ in resistance but which may individually swing over a wide resistance range during temperature cycling.

As the temperature was repeatedly cycled, the temperature, the applied voltage E, the voltage across the sample V_x, the voltage across the variable resistor V_h, and the variable resistance R_h were measured. In some cases, the voltage across the sample was kept constant. In other cases, the voltage across the sample was varied.

To begin with, both the resistance of the sample when it was biased positively, R^+, and the resistance of the sample when it was biased negatively, R^-, were calculated using the following formulas:

$$R^+ = (V_x^+/V_h^+)R_h \quad , \tag{15}$$

$$R^- = (V_x^-/V_h^-)R_h \quad . \tag{16}$$

R^+ and R^- turned out to be different. The results for a typical run are shown in Fig. 8. As can be seen, R^+ starts differing from R^- at the switching threshold temperature.

Table 2. Characteristics of the SiO-Ag samples that exhibited the thermal switching phenomenon.

Sample	Number of Layer Pairs	Layer Thickness SiO (Å)	Ag (Å)	Identity Period (A)	Substrate	Original Room Temperature Resistance (Ω)
1-(80)	264	24	22	47	Gold on Glass	1.4×10^7
2-(81)	108	59	55	114	Gold on Glass	1.2×10^8
3-(83)	59	89	84	117	Gold on Glass	6.3×10^8

The determined discrepancy between R^+ and R^- can be explained by assuming that the sample is generating a temperature dependent voltage ε as indicated in the equivalent circuit diagram Fig. 9. ε can be calculated from,

$$\varepsilon = \frac{1}{2}\left[(V^+_x - V^-_x) - (V^+_h - V^-_h)\frac{V^+_x + V^-_x}{V^+_h + V^-_h}\right] \tag{17}$$

The samples used in this experiment were made by evaporating alternate layers of silicon monoxide and silver on metallic and on glass substrates. Only 60% of the samples tried exhibited the switching phenomenon. But no correlation between the identity period and the switching properties could as yet be established.

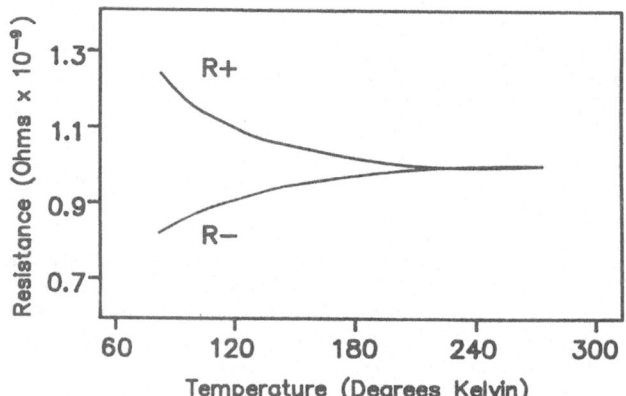

Fig. 8. Temperature dependence of R^+ and R^-. R^+ and R^- are the sample resistances when biased positively and negatively.

The characteristics of some of the samples found to be capable of thermal switching are summarized in Table 2.

Although multiple switching was also observed, single switching predominated. A representative switching process is shown in Fig. 11,. The resistance of the sample R_s was calculated using Eq. (10) and the two switching threshold temperatures $Th^{\downarrow}$ and $Th^{\uparrow}$ are indicated in this figure. These switching thresholds remained reasonably constant throughout the switching cycles as shown in Table 3. The switching characteristics of the samples examined were similar in nature and to avoid redundancy, the present description will deal with Sample I—(80) only because this sample was examined in greatest detail.

Table 3. Reproducibility of the switching threshold temperatures $Th^{\downarrow}$ and $Th^{\uparrow}$ for Sample 2-(80).

Run Number	1	2	3	4	5	6
$Th^{\downarrow}$ (K)	80	80	82	87	105	83
$Th^{\uparrow}$ (K)	149	149	144	144	137	135

Sample I—(80) had an original resistance of about 1.4×10^7 ohms at room temperature. Its area perpendicular to the current flow was approximately 1 cm^2. As the temperature was steadily lowered, the resistances increased moderately until the resistance started decreasing sharply to about 10^6 ohms at the temperature threshold $Th^{\downarrow}$ of 170 K. In this low resistance state the polarity of the voltage across the sample switched so that it was in opposition to the applied voltage, indicating that the thermally induced potential ε was opposite in polarity and greater than the IR—drop in the sample. This inverted potential ε, which was on the order of a few millivolts, persisted when the sample was disconnected from the circuit. The dependence of ε on temperature is shown in Fig. 10. When the sample was in this low resistance state, it remained in this state as long as the temperature was held constant.

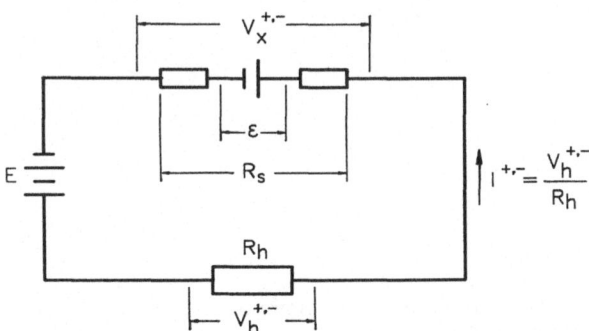

Fig. 9.   The equivalent circuit diagram during switching. ε
is the voltage generated by the sample during
temperature induced switching.

When the temperature was increased again, the sample switched into a higher state at the threshold temperature Th$^\uparrow$ of 215 K. Reaching room temperature again, the resistance of the sample measured higher (about $1.6 \times 10^7 \, \Omega$) than the initial resistance of the sample ($1.4 \times 10^7 \, \Omega$).

If the sample was then cooled down again after attaining this higher resistance state, the sample again switched into the low resistance state near the previous temperature Th$^\downarrow$. Cycling between the high and the low conductivity states was repeated several times with almost identical results as is shown in Table 3. The higher resistance at room temperature at the end of the cycle decreased as a function of time until the resistance reached the initial resistance at room temperature. In successive runs the same type of pattern was repeated. This behavior is depicted in Fig. 11a.

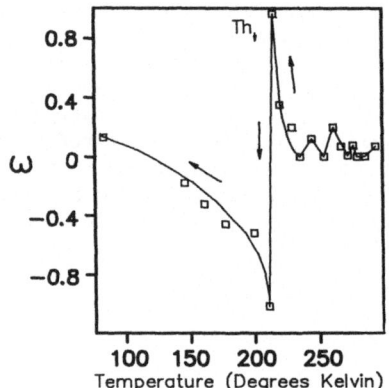

Fig. 10. Temperature dependence of the voltage generated by the sample during temperature induced switching.

After undergoing many cycles, however, the character of Sample 80 changed. The room temperature resistance of the sample eventually changed to less than 2 ohms and then 1.2 ohms, showing a deterioration of the switching characteristics. Before the samples deteriorated seriously, however, both switching thresholds Th$^\uparrow$ and Th$^\downarrow$ increased. This is shown in Fig. 11b where Th$^\downarrow$ moved to 215 K (Th$^\downarrow$ = 170 K in Fig. 11a) and Th$^\uparrow$ moved above 260 K.

Very indicative in Fig. 11b is an oscillation in the resistance as a function of temperature before the threshold temperature Th$^\downarrow$ is reached. This is reflected in the temperature behavior of ε which is shown in Fig. 10. Again the onset of the switching is preceded by oscillations of $\varepsilon(T)$ and Fig. 10 clearly shows that ε reverses polarity through Th$^\downarrow$ approaching normal conduction ($\varepsilon = 0$) at lower temperatures.

The preceding description clearly indicates that the type of switching exhibited by the SiO—Ag layered structures is unlike other types of switching which have to date been reported in literature. The behavior of the resistance, which decreases as temperature is decreased tends to go contrary to other switching results (with the possible exception of the high conductivity pulses in TTF—TCNQ as mentioned previously).

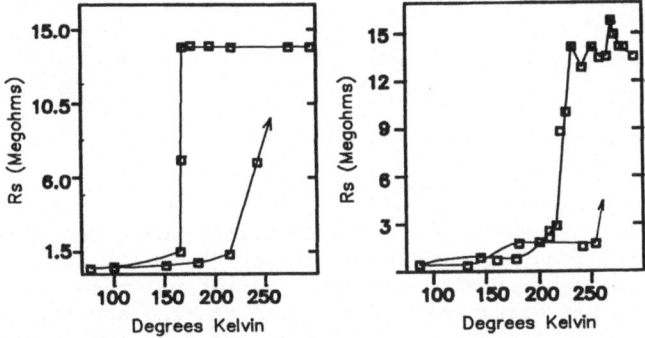

Fig. 11.   Resistance switching characteristic of a freshly prepared sample is shown in (a). The characteristic of a sample that is not far from failing after many switching cycles is shown in (b).

3.   Optical Properties

The optical properties of Ag—SiO superlattice stacks were described inclusively in previous publications by Abuel—Haija, McMarr and Madjid, (Abuel—Haija, McMarr and Madjid, 1979; Madjid and Abuel—Haija, 1980), and these results will consequently only be summarized in this section.

A transfer matrix technique was used by these authors to derive expressions for the transmissivity and reflectivity of these superlattice stacks by, first, representing each primary layer by its characteristic matrix containing all geometric and optical parameters of the layer. The characteristic matrix of the entire stack was subsequently obtained by forming the produce over all layers. In doing this, account was taken of the substrate, the ambience and also of the proper termination of the stack. The procedure was somewhat tedious, but straightforward. The final stack matrix contained the sum of three terms. The first term was the characteristic matrix of the ideal stack. The second term was a correction matrix accounting for the presence of interfacial boundaries, and the third represented a correction that took account of possible random variations in the layer thicknesses.

The results of such a calculation are compared to measure values in Fig. 12a for a stack structure shown in Fig. 12b. The experimental values refer to a stack of identity period 110 Å with the following layer thickness fractions, Ag:0.55; SiO:0.45. The theoretical calculations refer to the same identity period, but apportion the thickness fractions as follows, Ag:0.45; SiO:0.35; Interfacials:0.10. The primary layers, furthermore, are assumed to be subject to a random variation of ±10%.

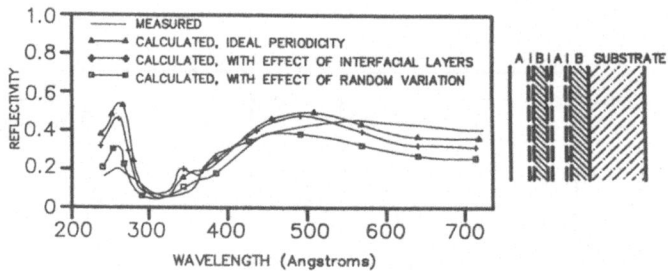

Fig. 12. Computational example. The reflectivity for a four-identity period silver-silicon monoxide superlattice (h = 110Å, A:Ag = 60 Å, B:SiO= 50 Å) is compared with computed theoretical results. The latter show the characteristic of an ideal staking and the effects of interfacial layers (I: 10% of h,n, and k of the interfacial layer assumed to be the average of the respective Ag and SiO values) and of random variation in identity periods {h = 110[1+(R/100)], −10 ≤ R≤ 10}.

Recently Abuel–Haija and Haddad (1984) subjected Ag:SiO superlattice stacks to external electric fields and ascertained a moderate effect on the transmissivity. These authors also remeasured the optical properties of seven year old samples and found relatively little change over this span of time (old samples, however, showed some fine structure in the transmissivity spectrum which was not evident in freshly prepared samples). This shows that Ag:SiO superlattices are surprisingly stable.

CONCLUSION

Although most of the recent work pertaining to superlattices focused on carefully preserving crystalline order by making use of the restriction that epitaxy presents, our own efforts concentrated on arrays of rather disordered layers which proved to exhibit a much greater range of puzzling, novel and potentially useful phenomena.

We have presented a summary of our experimental experiences with such arrays and wish to dedicate this work to Heinz Henisch who has been a friend to one of us over the last quarter century and a respected colleague and teacher to others. Heinz belongs to the vanishing breed of scholars of culture and gentlemanly decency. We wish him well. May his years be many and may they continue to be rich.

REFERENCES

Abuel—Haija, A. J. and F. H. Haddad, private communication (1984).

Abuel—Haiji, A. J., P. J. McMarr and A. H. Madjid, Appl. Opt. 18, 3123 (1979).

Allam, D. S. and K. E. G. Pitt, Thin Solid Films 1, 245 (1967).

Ball, C. J., "An Introduction to the Theory of Diffraction," Pergamon, New York (1971).

Boiko, B. T., A. N. Synelnikov and V. R. Kopach, "Proceedings of the International Conference on the Physics and Chemistry of Semiconductor Heterojunctions and Layer Structures," G. Szigeti, Ed., Akademiai Kiado, Budapest (1971).

Dinklage, J. and R. Frerichs, J. Appl. Phys. 34, 2633 (1963).

Doremus, R. H., J. Appl. Phys. 37, 2775 (1966).

DuMond, J. W. M. and J. P. Youtz, Phys. Rev. 48, 703 (1935).

Hartman, T. E., J. Appl. Phys. 34, 943 (1963).

Howson, R. P. and A. Taylor, Thin Solid Films 9, 109 (1971).

Kratky, O., Z. Elektrochem. 58, 49 (1954).

Madjid, A. H and A. J. Abuel—Haija, Appl. Opt. 19, 2612 (1980).

Maxwell—Garnett, J. C., Philos. Trans. R. Soc. (London) 203, 385 (1904); 205, 237 (1906).

Santana, P. H. A. and A. Rosato, Am. J. Phys. 41, 1138 (1973).

Sennett, R. S. and G. D. Scott, J. Opt. Soc. Amer. 40, 203 (1950).

Tolansky, S., "Multiple—Beam Interferometry of Surfaces and Films," Oxford (1948).

Tolansky, S., "Surface Microphotography," Interscience, New York (1960).

Tolansky, S., "Microstructures of Surfaces Using Interferometry," Elsevier, New York (1968).

Wagner, K. W., "Die Isolierstoffe der Elektrotechnik,," H. Schering, Ed., Springer, Berlin (1924).

White, E. W. and R. Roy, Solid State Communications 2, 151 (1964).

PIEZORESISTIVITY IN SEMICONDUCTING FERROELECTRICS

Ahmed Amin

Advanced Development Laboratory
Texas Instruments Incorporated
Attleboro, Massachusetts

ABSTRACT

The piezoresistive effect in semiconducting polycrystalline barium titanate and its solid solutions with lead and strontium titanate under different elastic and thermal boundary conditions will be reviewed. An account of this phenomenon based upon recent models of ferroelectricity and grain boundary potential is given. A comparison to silicon and germanium is attempted.

INTRODUCTION

The piezoresistive effect, i.e., the specific change of electrical resistivity of semiconductors with stress, is of considerable interest both from the fundamental (Herring, 1955) and device (sensor) technology viewpoints (Zaima et al., 1986). Many pressure, torque, vibration, and acceleration sensors which utilize this phenomenon are commercially available. These sensors are commonly fabricated on a precisely micromachined and etched n-type silicon wafer (diaphragm). The optimum diaphragm design is achieved by finite element techniques (Yasukawa et al., 1982).

In order to maximize the piezoresistive response with respect to incoming mechanical signals, four p-type resistors are etched in a definite pattern with respect to the crystallographic axes of the wafer and connected to form a Wheatstone bridge using standard integrated circuit technology. Piezoresistivity in junctionless silicon on sapphire SOS (Hynecek, 1974), n-channel inversion layers MOSFET's fabricated on

201

SOS film (Zaima et al., 1986), and polycrystalline silicon (French et al., 1985) are recent examples of the continued effort in piezoresistive silicon technology.

In addition to silicon and germanium, there has been an increasing interest in materials which exhibit large hydrostatic piezoresistive coefficient. For silicon and germanium, the longitudinal Π_{11} and transverse Π_{12} piezoresistive coefficients are quite large when compared to the hydrostatic coefficient. Typical values of these coefficients for p-type silicon in the <111> direction (where coefficients are maxima), are +90 ($\times 10^{-11}$ m^2/N) and −42 ($\times 10^{-11}$ m^2/N) respectively. The hydrostatic piezoresistive coefficient, however, is a linear combination of these two coefficients (= 90 x 10^{-11}− 2 x42 x10^{-11}= +6 x10^{-11} m^2/N).

Piezoresistive hydrostatic sensing, particularly for high pressures, would provide an attractive alternative to the silicon diaphragm devices which have limited strength. Moreover, pressure studies of electrical resistivity in polycrystalline semiconducting barium titanate would enhance understanding of the positive temperature coefficient of resistance (PTCR), and other grain boundary related phenomena which are observed in this class of materials.

We will begin by reviewing the semiconducting properties of polycrystalline donor doped barium titanate BaTiO$_3$, and the positive temperature coefficient of resistance (PTCR) anomaly observed above the ferroelectric-paraelectric Curie temperature T$_c$. Models of the origin of interfacial acceptor states, that lead to the formation of grain boundary barriers, will be briefly discussed. Phenomenological description of piezoresistivity and piezoresistance measurement results will be presented in some details. The interaction of mechanical stress with ferroelectric properties of the grain boundary will be explained in terms of the recent models of ferroelectricity. Finally, a comparison to silicon and germanium will be given.

SEMICONDUCTING BARIUM TITANATE

Single crystal ferroelectric barium titanate BaTiO$_3$ belongs to the ferroic species m3mF4mm. The term ferroic was originally introduced by Aizu (1972, 1973) to describe the many types of mimetically twined crystals in which the orientation of one or more twin components may be affected by the application of a suitably chosen driving force/field. Ferroic crystals can be classified into symmetry species according to their high-and-low temperature symmetries. The symmetry species symbols consist of the high temperature point group followed by the low temperature point group. The two groups are separated by the letter F indicating a ferroic behavior in the low temperature phase.

Ferroelectrics, ferromagnetics, and ferroelastics are examples of ferroic crystals in which the orientation (domain) states could be switched by the application of electric, magnetic, or elastic fields respectively. The well-known hysteresis phenomenon between conjugate pairs (e.g., electric displacement D and applied electric field E in ferroelectrics) is among the characteristics of this class of materials.

Furthermore, ferroic crystals may be classified according to the nature of the property tensor of the orientation states as primary, secondary, and tertiary, or even higher order (Amin et al., 1980). In the absence of a strong electric field, polycrystalline barium titanate belongs to the symmetry species $\infty/\infty/mmm$. The paraelectric (Pm3m) perovskite structure of barium titanate above its Curie temperature T_c (120° C) consists of corner shared TiO_6 octahedra with the Ti atoms placed at the corners of the cube and the Ba atom at the center of the body diagonal.

Semiconducting, donor-doped polycrystalline barium titanate exhibits an anomalous increase in resistivity (Fig. 1) known as positive temperature coefficient of resistivity (PTCR) above the ferroelectric Curie temperature T_c (Saburi, 1959; Heywang 1961). Goodman (1963) compared the temperature dependence of d.c. resistivity of semiconducting polycrystalline $BaTiO_3$ doped with 0.05% Sm to that of a single crystal barium titanate of the same composition. No appreciable anomaly in resistivity above T_c was observed in single crystal $BaTiO_3$.

PTCR materials are based on the crystalline solution $(A_y Ba_{1-y-x} D_x)TiO_3$, where A may be any of the divalent Curie temperature shifters such as Sr^{2+}, or Pb^{2+}, or grain growth inhibitors such as Ca^{2+}, and $y > 0$. The Curie temperature of barium titanate is decreased linearly with Sr^{2+} substitution for Ba^{2+}, the Curie temperature of $SrTiO_3$ is near 0°K. Complete solid solution occurs between $BaTiO_3$ and $PbTiO_3$, substitution of Pb^{2+} for Ba^{2+} has the effect of raising the Curie temperature monotonically towards that of $PbTiO_3$ (490°C).

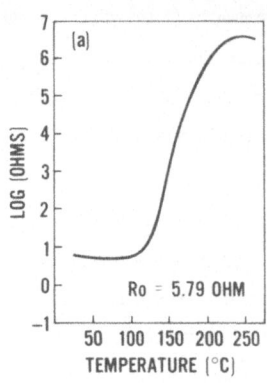
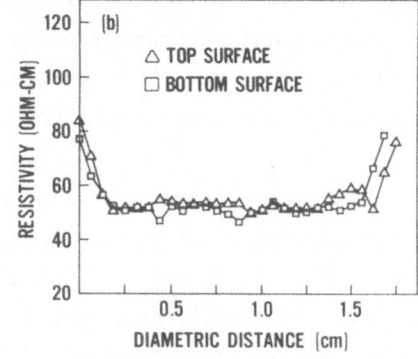

Fig. 1. (a) Resistance-temperature characteristics for Yttrium doped (Ba, Ca) TiO$_3$ and (b) Resistance profile using four-point probe.

Semiconducting properties are produced by (i) doping with a trivalent element (e.g., La^{3+}, Y^{3+},...etc) which substitutes for Ba^{2+} on the lattice site; or (ii) doping with a pentavalent element (e.g. Nb^{5+}, Sb^{5+},...etc) which substitutes for titanium. In this manner free electrons are generated in the titanium 3d conduction band through the formation of Ti^{4+} Ti^{3+} complex as follows, (Ba$^{2+}_{1-x}$ D$^{3+}_x$) (Ti$^{4+}_{1-x}$ Ti$^{3+}_x$) O$_3$. The dopant concentration x is usually within 0.3 atom percent, depending upon the dopant type and the required electrical properties of the final composition.

Development of large (6 to 8 orders of magnitude) PTC anomaly, requires low level (approximately 500 ppm) acceptor doping with ions such as Mn. The magnitude of PTC anomaly is field sensitive, it decreases with increasing applied voltage. Commercial PTC resistors have room temperature resistivities between 20 ohm-cm and 20K ohm-cm and Curie temperatures between -80°C and 300°C.

A plausible explanation of the PTC effect is given by Heywang (1961) as being due to the rapid decrease in dielectric constant above the ferroelectric-paraelectric Curie temperature which leads to a sharp rise

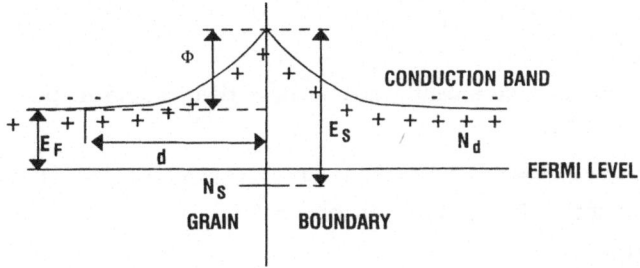

Fig. 2.   Band diagram for grain boundary (Heywang, 1961).

in the height of the grain boundary potential barrier. In Heywang's treatment the grain boundary is modeled as a strong depletion layer, back-to-back Schottky barrier (Fig. 2).

Following Heywang (1961, 1963, 1964, 1971), the expression for the static, quiescent condition (zero current) barrier height ϕ is obtained by solving the one dimensional Poisson's equation under the assumption that barium titanate is a linear and isotropic dielectric

$$\phi = (e^2 N_d d^2)/(2\epsilon\epsilon_0) \tag{1}$$

where e is the electronic charge, N_d the donor concentration per unit volume, d the thickness of the depletion layer, ϵ the relative permittivity (dielectric constant) and ϵ_0 the free space permittivity.

The charge in the surface states must be equal in magnitude to the charge in the depletion region (charge neutrality)

$$n_s = 2N_d d \tag{2}$$
$$n_s = n_{a,s} - n_{d,s} \tag{3}$$

where n_s is the net concentration of ionized surface states, $n_{a,s}$ the total concentration of ionized surface acceptor states (chemidsorbed oxygen and cationic vacancies), $n_{d,s}$ the concentration of surface donor states (considered negligible). Substituting from Eq. (2) in Eq. (1), the barrier height takes the form

$$\phi = (e^2 n_s^2)/(8\epsilon\epsilon_0 N_d) \tag{4}$$

In Heywang's treatment the surface states are assumed to lie at one fixed energy level E_s, well below the Fermi level (fully occupied), and the donor levels are sufficiently close to the conduction band (fully ionized), n_s is given by

$$n_s = N_s/[1 + \exp\ (E_f + \phi - E_s)/kT)] \tag{5}$$

the resistivity ρ is given by

$$\rho = \rho_0 \exp (\phi/kT) \tag{6}$$

where k is Boltzmann's constant, and ϕ is a mean barrier potential.

Above the Curie temperature T_c, the Curie–Weiss law is obeyed, the dielectric constant ε for BaTiO$_3$ is given by

$$\varepsilon = C/(T-T_0) \tag{7}$$

where T_0 is the extrapolated Curie–Weiss temperature (381 K), C the Curie constant (1.7×10^5 K). Therefore, the rapid decrease in dielectric constant above the ferroelectric transition temperature T_c is responsible for the sharp rise in barrier height, hence the observed PTC anomaly.

The voltage dependence of the resistivity is attributed to Schottky emission across the barrier (Mallick et al., 1968). The following expression is given by Heywang (1964)

$$\phi = \phi_0 [1-(eU/4\phi_0)]^2 \tag{8}$$

where ϕ_0 is given by Eq. (1) and U is the applied voltage such that $eU<4\phi_0$. In deriving Eq. (8) the entire voltage drop was assumed to be across the barrier between two grains and the occupation of the surface states is not changed due to the applied voltage. A comprehensive analysis of barriers in semiconductors is given by Henisch (1984).

Heywang's model, however, requires the assumption of a high effective dielectric constant ε(eff) below the Curie temperature to account for the lower and relatively constant resistivity observed in the ferroelectric state. The value of ε (eff) corresponds to that measured with large signal. Jonker (1964) has refined the model by suggesting that the normal components of spontaneous polarization (ΔP_n) at the end of alternate 90° domains could effectively compensate the surface states present at the grain boundaries, therefore, eliminating a potential barrier there. Thus, for temperatures below T_c, an approximate expression for the barrier height could be written as

$$\phi \sim [e^2 n_s^2 - (\Delta P_n)^2]/8\varepsilon\varepsilon_0 N_d \tag{9}$$

A method based on Devonshire's formalism was developed by Kulwicki et al. (1970) to calculate the potential profiles of depletion layers in semiconducting, single domain ferroelectric PTC material. Below the Curie point, the calculations were in agreement with the resistivity-temperature behavior of doped barium titanate. The model, however, did not predict a large resistance discontinuity at the Curie point. A possible mechanism for this discrepancy was attributed to lattice deformation and grain boundary clamping below T_c, which would tend to lower barrier height. Above T_c, however, the clamping disappears, leading to a higher barrier potential.

It is generally believed that the grain boundary barrier formation in semiconducting barium titanate is due to chemisorbed oxygen, or halogen (Jonker, 1964, 1967). Acceptor states are generated at the grain boundary due to oxygen adsorption and diffusion during sintering. Conduction electrons become trappped in the interfacial states, leading to the formation of depletion layers. Since a large amount of charge is trapped at the grain boundary, a high capacitance is expected. In this model, acceptor Mn sites would generate lower trapping states.

Other mechanisms have been proposed for the nature of the interfacial acceptor states. These include oxidized impurities (Kahn, 1971) which are expected to provide fairly deep traps. An alternative mechanism, gradient of Ba-vacancies in the form of thin skin on the surface of each grain has been proposed by Daniels et al. (1979). A gradient of Ti-vacancies in the grain surface region has also been suggested (Lewis et al., 1985). At high donor concentrations, the primary compensating defect is the titanium vacancy (Chan et al., 1986). More recently, Scholl (1986) developed a model for potential barrier height at grain boundary in polycrystalline semiconductors. The barrier height can fluctuate along a grain boundary due to different local curvatures, being lowest at corners of the grains.

PIEZORESISTIVITY-PHENOMENOLOGICAL DESCRIPTION

The piezoresistive effect in single crystal semiconductors can be described in the following manner. Following Mason (1957), the electric field E_i is expressed in terms of the current density I_j and applied stress X_{kl} as follows

$$E_i = E_i (I_j, X_{kl}) \qquad i,j,k,l = 1,2,3 \qquad (10)$$

In what follows the summation convention over repeated indices in the same term is implied. Expanding in a McLaurin's series about the origin (state of zero current and stress)

$$
\begin{aligned}
dE_i = & (\partial E_i / \partial I_j)\, dI_j + (\partial E_i / \partial X_{kl})\, dX_{kl} \\
& + 1/2! \, [(\partial^2 E_i / \partial I_j \partial I_m)\, dI_j\, dI_m + (\partial^2 E_i / \partial X_{kl} \partial X_{no}) dX_{kl}\, dX_{no} \\
& + 2\, (\partial^2 E_i / \partial X_{kl} \partial I_j)\, dX_{kl}\, dI_j] + \dots \text{ H.O.T} \qquad (11)
\end{aligned}
$$

the expansion coefficient in Eq. (11) have the following meanings

$(\partial E_i/\partial I_j) = \rho_{ij}$ (electric resistivity tensor)

$(\partial E_i/\partial X_{kl}) = d_{ikl}$ (converse piezoelectric tensor)

$(\partial^2 E_i/\partial X_{kl} \partial I_j) = \partial/\partial X_{kl} \, (\partial E_i/\partial I_j) = \Pi_{ijkl}$

                          (piezoresistivity tensor)

$(\partial^2 E_i/\partial I_j \partial I_m) = \rho_{ijm}$ (non-linear resistivity tensor)

$(\partial^2 E_i/\partial X_{kl} \partial X_{no}) = \delta_{iklno}$ (non-linear piezoelectric tensor)

replacing the differentials in Eq. (11) by the components themselves

$$E_i = \rho_{ij} I_j + d_{ikl} X_{kl} + 1/2 \, \rho_{ijm} I_j I_m + 1/2 \, \delta_{iklmo} X_{kl} X_{no}$$
$$+ \Pi_{ijkl} X_{kl} I_j \tag{12}$$

The transformation law of Π_{ijkl} (a fourth rank polar tensor) is as follows

$$\Pi'_{ijkl} = (\partial x'_i/\partial x_m)(\partial x'_j/\partial x_n)(\partial x'_k/\partial x_o)(\partial x'_l/\partial x_p) \Pi_{mnop} \tag{13}$$

where the primed and unprimed components refer to the new and old coordinate systems respectively, and the determinant $||\partial x'_i/\partial x_m||, \dots$ etc are the Jacobian of the transformation. A general fourth rank tensor has 81 independent components however, the piezoresistivity tensor Π_{ijkl} has the following internal symmetry

$$\Pi_{ijkl} = \Pi_{ijlk} = \Pi_{jilk} \tag{14}$$

which reduces the number of independent tensor coefficients to 36. The point group symmetry of the crystal imposes further constraints on the remaining 36 coefficients. If the crystal contains an inversion symmetry operation, all odd rank tensor coefficients will vanish and Eq. (12) takes the form

$$E_i = \rho_{ij} I_j + \Pi_{ijkl} X_{kl} I_j \tag{15}$$

rearranging Eq. (15)

$$(E_i(X) - \rho_{ij}(0) I_j)/I_j = \rho_{ij}(X) - \rho_{ij}(0) = \Pi_{ijkl} X_{kl} \tag{16}$$

thus

$$\delta\rho_{ij} = \Pi_{ijkl} X_{kl} \tag{17}$$

Eq.(17) could be written in terms of the strain conjugate x_{op} as follows

$$\delta\rho_{ij} = M_{ijop} x_{op} \tag{18}$$

where M_{ijop} is the elastoresistance tensor. The elastoresistance tensor is related to the piezoresistance tensor by the relationship

$$M_{ijop} = \Pi_{ijkl} c_{klop} \tag{19}$$

where c_{klop} is the elastic stiffness tensor. Equation (12) contains the various coupling coefficients which contribute to the resistivity change of a non-centric single crystal under applied stress. These are the converse piezoelectric, the non-linear resistivity, and the nonlinear piezoelectric coefficients. In a centric crystal, as pointed out earlier, the only intrinsic contribution to the resistivity change under

represent the piezoresistivity tensor in the reduced notation as follows:
applied stress stems from the piezoresistive term. It is convenient to

$$\Pi_{ijkl} = \Pi_{ij} \qquad (i,j = 1,2,\ldots,6) \qquad\qquad (20)$$

The non vanishing tensor coefficients for each point group are
obtained by applying the generating elements of the point group to the
piezoresistivity tensor -then applying the constraints imposed by
Neumann's law (Nye, 1976; Bhagavantam, 1966). The number of independent
coefficients required to completely specify the piezoresistivity tensor
in barium titanate above and below its Curie temperature T_c are, six
$(\Pi_{11}, \Pi_{12}, \Pi_{13}, \Pi_{33}, \Pi_{44}, \Pi_{66})$ for the tetragonal point group 4mm, and
three $(\Pi_{11}, \Pi_{12}, \Pi_{44}$ for the point group m3m. The full piezoresistive
matrices for the above symmetries are

(i) Tetragonal (4mm)

$$
\begin{bmatrix}
\Pi_{11} & \Pi_{12} & \Pi_{13} & 0 & 0 & 0 \\
 & \Pi_{11} & \Pi_{13} & 0 & 0 & 0 \\
 & & \Pi_{33} & 0 & 0 & 0 \\
 & & & \Pi_{44} & 0 & 0 \\
 & & & & \Pi_{44} & 0 \\
 & & & & & \Pi_{66}
\end{bmatrix}
$$

(ii) Cubic (m3m)

$$
\begin{bmatrix}
\Pi_{11} & \Pi_{12} & \Pi_{12} & 0 & 0 & 0 \\
 & \Pi_{11} & \Pi_{12} & 0 & 0 & 0 \\
 & & \Pi_{11} & 0 & 0 & 0 \\
 & & & \Pi_{44} & 0 & 0 \\
 & & & & \Pi_{44} & 0 \\
 & & & & & \Pi_{44}
\end{bmatrix}
$$

Polycrystalline semiconducting barium titanate, however, belongs to
the isotropic symmetry species ∞/∞/mmm (a centrosymmetric group). In
the absence of strong electric fields, this symmetry species remains
invariant upon heating (from below) or cooling (from above) the Curie
temperature T_c, with only two independent coefficients (Π_{11}, Π_{12})
required to completely specify the piezoresistive tensor. The third Π_{44}
coefficient which is required by symmetry is a linear combination of the
longitudinal and transverse components $[2(\Pi_{11}-\Pi_{12})]$. The matrix
representation of the piezoresistive tensor for this species is

(iii) Isotropic ($\infty/\infty/mmm$)

$$
\begin{bmatrix}
\Pi_{11} & \Pi_{12} & \Pi_{12} & 0 & 0 & 0 \\
 & \Pi_{11} & \Pi_{12} & 0 & 0 & 0 \\
 & & \Pi_{11} & 0 & 0 & 0 \\
 & & & \Pi_{44} & 0 & 0 \\
 & & & & \Pi_{44} & 0 \\
 & & & & & \Pi_{44}
\end{bmatrix}
$$

It can be shown that the piezoresistance coefficient $1/R_0 (\partial R/\partial X)$ and the piezoresistivity coefficient $1/\rho_0 (\partial\rho/\partial X)$ are connected by the following relations

(i) Hydrostatic pressure

$$X_1 = X_2 = X_3 = p, \; X_4 = X5 = X_6 = 0$$

$$1/R_0 (\partial R/\partial p) - (s_{11}+2s_{12}) = 1/\rho_0 (\partial\rho/\partial p) = (1/\rho_0)\Pi_h \tag{21}$$

(ii) Uniaxial stress parallel to current flow

$$1/R_0 (\partial R/\partial X) + (s_{11}-2s_{12}) = 1/\rho_0 (\partial\rho/\partial X) = (1/\rho_0)\Pi_{11} \tag{22}$$

(iii) Uniaxial stress perpendicular to current flow

$$1/R_0 (\partial R/\partial X) - s_{11} = 1/\rho_0 (\partial\rho/\partial X) = (1/\rho_0)\Pi_{12} \tag{23}$$

where $X_1 (=X_{11})$, $X_2(=X_{22})$, $X3 (=X_{33})$ are the normal stress components, $X_4 (=X_{23})$, $X_5 (=X_{13})$, $X_6 (=X_{12})$ the shear components, $\Pi_h (=\Pi_{11}+2\Pi_{12})$ the hydrostatic piezoresistive coefficient, s_{11} and s_{12} the elastic compliances (in reduced notation). For semiconducting barium titanate, under hydrostatic pressures, the difference between the two coefficients is about 0.2% at room temperature (Amin, 1987).

REVIEW OF EXPERIMENTAL RESULTS

Semiconducting barium titanate PTC ceramics are usually prepared from the starting carbonates or oxides by a solid state reaction technique. Metallization is usually accomplished by using silver or aluminum. Details for preparing metallized ceramic PTC elements are given by Amin et al. (1983).

Sauer et al. (1959) were perhaps the first to report a large piezoresistance response to hydrostatic pressure for stoichiometric compositions in the system barium strontium lanthanum titanate at room temperature. The resistivity was found to increase non-linearly with pressure up to approximately 550 Mpa. The maximum hydrostatic piezoresistance coefficient Π_h was 700×10^{-11} m^2/N, approximately two orders of magnitude larger than silicon. The resistance of the sample was found to vary linearly under uniaxial stress up to a stress level of approximately 7 MPa. Π_{11} was positive for compressive stresses and negative for tensile, and was about two to three times larger than that of silicon.

Saburi (1960) measured the temperature dependency of both longitudinal Π_{11} and transverse Π_{12} piezoresistivity coefficients for a series of compositions in the system barium strontium cerium titanate under compressive stresses. Far below the Curie point, both Π_{11} and Π_{12} were small and showed a negative peak around the Curie point, which depended upon the Sr content. Above the Curie point, the piezoresistivity decreased rapidly, changed sign and tended to a positive value asymptotically.

The effect of combined d.c. and a.c. stresses on the complex resistivity change of donor doped barium strontium titanate at different temperatures has been studied by Heywang et al. (1966). They concluded that near the Curie point the piezoresistivity coefficient becomes more negative with increasing d.c. bias. However, with no bias, the piezoresistive coefficient remained positive below and above the Curie point. Therefore, the manner in which the piezoresistive coefficients vary with temperature depends to a large extent on the elastic boundary conditions which exist near or at the grain boundary.

The temperature dependence of the hydrostatic piezoresistance coefficient of a high Curie temperature (175°C) composition (barium calcium lead yttrium titanate) has been studied by Amin (1986). The resistance vs hydrostatic pressure measurements were taken below the Curie temperature, over the pressure range 0 to 35 MPa and the temperature range 25°C to 130°C. This composition exhibited a linear resistance change with hydrostatic pressure. The hydrostatic piezoresistance coefficient Π_h was negative below 50°C, and changed sign above that temperature. Π_h increased rapidly on approaching the Curie temperature, with a value at 100°C two orders of magnitude larger than that of silicon.

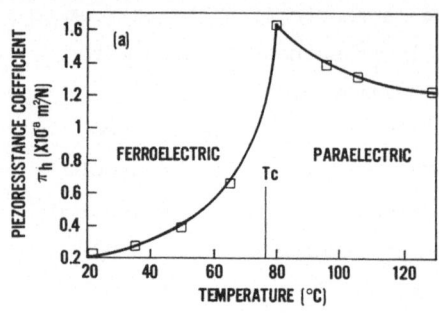 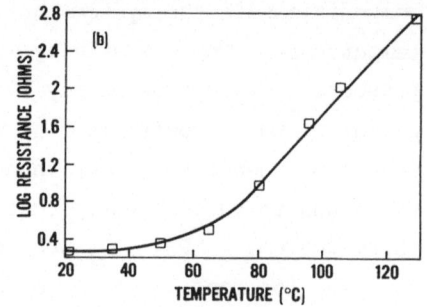

Fig. 3.   (a)   Hydrostatic piezoresistance coefficient vs temperature
for semiconducting (Ba, Sr) TiO$_3$ and (b) Temperature
dependence of R$_0$ (resistance at zero hydrostatic
pressure) for semiconducting (Ba, Sr) TiO$_3$ (Amin, 1987).

Figure 3(a) shows the temperature dependence of the hydrostatic
piezoresistance coefficient of an yttrium doped barium strontium titanate
composition below and above its ferroelectric-paraelectric transition
temperature (Amin, 1987).   In this system the resistance increased
linearly with hydrostatic pressures (0–35 MPa) with correlation
coefficient 0.995 at all temperatures of interest.   Figure 3(b) depicts
the manner in which the resistance at zero pressures varies with
temperature.

Igarashi et al. (1985) have measured the longitudinal and transverse
piezoresistance coefficients for two barium strontium titanate
compositions with the same Curie temperature (65°C) and different
resistivity-temperature characteristics.   The measurements were conducted
over the pressure range (0–140 MPa) and the temperature range
(20°C–115°C).   At low stress levels the sign of piezoresistance
coefficient under compression was opposite to that under tension.   The
change of resistivity with stress was accompanied by hysteresis at high
stress levels.   This hysteresis effect was explained as being due to
plastic deformation, which is large under uniaxial stress and negligible
under hydrostatic pressure.

PHENOMENOLOGICAL DESCRIPTION OF THE STRESS DEPENDENCE OF BARRIER HEIGHT
        The ferroelectric nature of the grain boundary in semiconducting
barium titanate suggests phenomenological theories.   In many
ferroelectric crystals, it has often proven useful to correlate the
dielectric, piezoelectric, and elastic properties of paraelectric and

ferroelectric phases by a phenomenological thermodynamic theory. Therefore, the dielectric properties of the "insulating" grain boundary in semiconducting barium titanate could be analyzed under different thermal, elastic, and electric boundary conditions.

The change of the thermodynamic potential associated with the onset of ferroelectric phase is described through a Taylor series expansion in powers of the order parameter (e.g., dielectric polarization in simple proper ferroelectrics) and of the coupling parameters to other interesting thermodynamic variables. The permitted terms and coupling parameters in this series representation of the thermodynamic potential are limited by the crystal symmetry of the prototypic paraelectric phase, and the coefficients of these terms are assumed to be continuous through any of the phase transitions into and between ferroelectric phases.

For many crystals this Landau–Ginzburg–Devonshire formalism gives an excellent semiquantitative description of the dielectric, elastic, piezoelectric, and electrothermal properties when only the lowest order terms are linearly temperature dependent, higher order terms temperature independent and often for a very truncated series expansion (Devonshire, 1949, 1951). For a more accurate description over a wider temperature range, additional higher order terms, and/or temperature dependent higher order coefficients can be included (Buessem et al., 1966; Goswami et al., 1968).

Consider the free energy density function for a proper ferroelectric derived from a prototypic symmetry group Pm3m. For Brillouin zone center lattice modes, the Landau–Ginzburg–Devonshire free energy density may be written as a power series in dielectric polarization P_i (i=1,2,3) as follows

$$
\begin{aligned}
G = {}& \alpha_1(P_1{}^2+P_2{}^2+P_3{}^2)+\alpha_{11}(P_1{}^4+P_2{}^4+P_3{}^4)+\alpha_{12}(P_1{}^2P_2{}^2+P_2{}^2P_3{}^2+P_3{}^2P_1{}^2)\\
& +\alpha_{111}(P_1{}^6+P_2{}^6+P_3{}^6)+\alpha_{123}[P_1{}^4(P_2{}^2+P_3{}^2)+P_2{}^4(P_3{}^2+P_1{}^2)+P_3{}^4(P_1{}^2+P_2{}^2)]\\
& +\alpha_{123}P_1{}^2P_2{}^2P_3{}^2-1/2s_{11}(X_1{}^2+X_2{}^2+X_3{}^2)-s_{12}(X_1X_2+X_2X_3+X_3X_1)\\
& -1/2s_{44}(X_4{}^2+X_5{}^2+X_6{}^2)-Q_{11}(X_1P_1{}^2+X_2P_2{}^2+X_3P_3{}^2)\\
& -Q_{12}[X_1(P_2{}^2+P_3{}^2)+X_2(P_3{}^2+P_1{}^2)+X_3(P_1{}^2+P_2{}^2)]-Q_{44}(X_4P_2P_3+X_5P_3P_1+X_6P_1P_2)
\end{aligned}
$$
$$(24)$$

where α_i, α_{ij}, α_{ijk} (in reduced tensor notation) are the dielectric stiffness and high-order stiffness coefficients at constant stress Q_{11}, Q_{12}, Q_{44} are the electrostriction constants written in polarization notation. In Eq. (24) the tensile stresses are denoted by X_1, X_2, X_3, and the shear components by X_4, X_5, X_6 respectively. The expression is complete up to all six power terms in polarization, but contains only first order terms in electrostrictive and elastic behavior.

The first partial derivatives of the free energy density with respect to the components of P_i, X_i, and T give the conjugate parameters, the electric field E_i, the negative of the strain $-x_{ij}$, and the entropy change $-S$, respectively

$$(\partial G/\partial P_i) = E_i \tag{25}$$

$$(\partial G/\partial X_{ij}) = -x_{ij} \tag{26}$$

$$(\partial G/\partial T) = -S \tag{27}$$

Appropriate second partial derivatives give the dielectric stiffnesses $1/x_{ij}$ (i.e., inverse dielectric susceptibilities), elastic compliances s_{ijkl}, and piezoelectric constants b_{ijk}

$$(\partial^2 G/\partial P_i \partial P_j) = 1/x_{ij} \tag{28}$$

$$(\partial^2 G/\partial X_{ij} \partial X_{kl}) = -s_{ijkl} \tag{29}$$

$$(\partial^2 G/\partial P_i \partial X_{jk}) = -b_{ijk} \tag{30}$$

the dielectric constants ε_{ij} are equal to the dielectric susceptibilities x_{ij} augmented by 1.

Equation (25) with $E_i = 0$, and the condition that equations (26) and (27) must have positive values, give the stability conditions for the spontaneously polarized states. Under zero stresses [all X_i vanish in Eq. (24)], the solutions of Eq. (24) which are of interest include the paraelectric cubic (Pm3m) and ferroelectric tetragonal (P4mm) states

(i) Paraelectric state $(T > T_c)$

$$P^2_1 = P^2_2 = P^2_3 = 0; \; G = 0$$
$$x_{11} = x_{22} = x_{33} = x_p, \; x_{12} = x_{23} = x_{31} = 0$$
$$(1/x_p) = 2\alpha_1 \varepsilon_0 \tag{31}$$

(ii) Ferroelectric state $T < T_c)$
$$P^2_1 = P^2_2 = 0$$

$$P^2_3 = \frac{-\alpha_{11} + [(\alpha_{11})^2 - 3\alpha_1 \alpha_{111}]^{.5}}{3\alpha_{111}} \tag{32}$$

$$G = \alpha_1 P^2_3 + \alpha_{11} P^4_3 + \alpha_{111} P^6_3 \tag{33}$$

$$x_{11} = x_{22} \neq x_{33}, \ x_{12} = x_{23} = x_{31} = 0$$

$$(1/x_{33}) = (2\alpha_1 + 12\alpha_{11}P_3^2 + 30\alpha_{111}P_3^4)\epsilon 0 \tag{34}$$

$$(1/x_{11}) = (2\alpha_1 + 2\alpha_{12}P_3^2 + 2\alpha_{112}P_3^2)\epsilon 0 \tag{35}$$

where ϵ_0 $(8.854 \times 10^{-12}$ F/m) is the free space permittivity.

In ferroelectric polycrystalline solids, a complete free energy function is particularly valuable, since often in a polycrystalline ensemble the elastic and electric boundary conditions upon the individual crystallites are uncertain. It is frequently not clear whether "unusual" properties are intrinsic and to be associated with these boundary conditions, or are extrinsic and associated with such phenomena as domain and phase boundary motion.

A properly developed free energy density function will permit the manner in which these parameters change under different elastic and electric boundary conditions to be evaluated (Amin et al., 1985, 1986). It can be shown that for uniform tensile stresses, the α_1 coefficient in Equations (24), (32), (33), (34), and (25) can be written as

$$\alpha_1(X) = \alpha_1 + [Q_{12}(X_1 + X_2) + Q_{11}X_3] \tag{36}$$

where X_1, X_2, X_3 are the normal stress components. Therefore, the manner in which the dielectric constant and other tensor properties change with some postulated mechanical boundary conditions can be easily computed.

Under hydrostatic pressure conditions, $X_1 = X_2 = X_3 = p$, $X_4 = X_5 = X_6 = 0$, thus substituting for these values of X in Equation (36), the dielectric inverse susceptibility of the paraelectric state Eq. (31) becomes

$$\chi^{-1}_{p} = 2(\alpha_1 + Q_h p \epsilon_0^{-1})\epsilon_0 \tag{37}$$
$$= 2\beta[T - (T_0 - Q_h \beta^{-1}p)]\epsilon_0 \tag{38}$$

where, $Q_h (= Q_{11} + 2Q_{12})$ is the hydrostatic electrostriction coefficient, and $\beta = 1/(\epsilon_0 C)$ with C being the Curie-Weiss constant. Therefore, above T_c the inverse dielectric susceptibility follows a Curie-Weiss law with a pressure dependent Curie-Weiss temperature $T_0(p)$

$$T_0(p) = T_0(0) - Q_h \beta^{-1} \epsilon_0 p^{-1} \tag{39}$$

at constant temperature, the Curie-Weiss law can be expressed as follows

$$\chi^{-1} = 2Q_h(p - p_0) \tag{40}$$

where $p_0 = -\beta\epsilon_0[T - T_0(0)]/Q_h$.

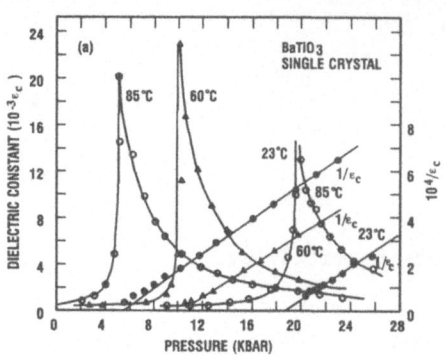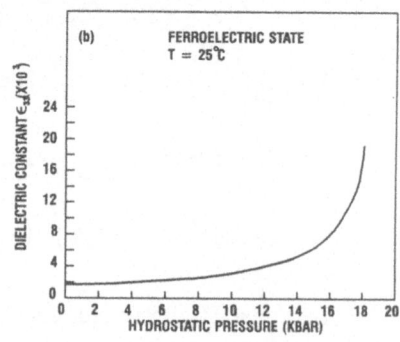

Fig. 4.　(a)　The effect of hydrostatic pressure on the dielectric constant of BaTiO$_3$ at various temperatures (Samara, 1966).

The effect of hydrostatic pressure on the ferroelectric properties of single crystal barium titanate has received considerable attention. Its results can be considered as typical for those transitions associated with Γ_{15} soft zone center optic phonons (Samara, 1970; Samara et al., 1975). Figure 4 (a) shows the variation of the dielectric constant ε_c (= ε_{33}) with hydrostatic pressure at various temperatures (Samara, 1966). Both the first order transition temperature (T_c) and the extrapolated Curie-Weiss temperature (T_0) decrease with increasing pressure.

However, the difference (T_c-T_0) tends to diminish with increasing pressure. This suggests a tendency towards second order characteristics at high pressure. Figure 4(b) depicts the hydrostatic pressure (0 to 18 kbar) dependence of the ferroelectric state dielectric constant ε_{33} using phenomenological expressions and the free energy coefficients of Buessem et al. (1966). Figure 5 depicts phenomenological predictions of the lower range 0 to 100 MPa (i.e., 0 to 1 Kbar) hydrostatic pressure dependence of the ferroelectric state dielectric constants ε_{33}, ε_{11}.

The free energy tensor coefficient under zero stresses for single crystal barium titanate (Buessem et al., 1966) in MKS units are, α_1= $0.33(T-T_0) \times 10^6$ (Vm/C), α_{11}= $(-20.25+0.47(T-T_c) \times 10^7$ (Vm^5/C^3), α_{111}= $(27-0.54(T-T_c)) \times 10^8$ (Vm^9/C^5), α_{12}= 3.6×10^8 (Vm^5/C^3), α_{112}= 4.5×10^9 (Vm^9/C^5), s_{11}= 8.7×10^{-12} (N/m^2), s_{12}=3.35×10^{-12} (N/m^2), s_{44}= 8.89×10^{-12} (N/m^2), Q_{11}= 0.1107 (m^4/C^2), Q_{12}= -0.0432 (m^4/C^2), Q_{44}=0.0652 (m^4/C^2), T_c= 120 ($^\circ$C), T_0=108 ($^\circ$C).

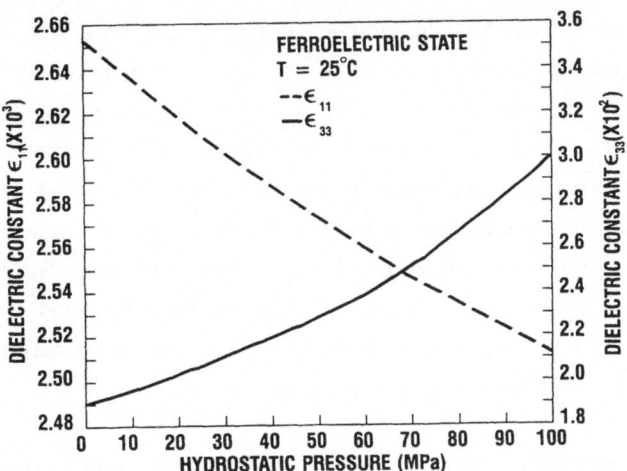

Fig. 5.   Hydrostatic pressure dependence of BaTiO$_3$ dielectric constants ε_{11} and ε_{33} at room temperature.

In terms of Cochran (1961) soft optical phonon theory, the strong pressure dependence of the dielectric consant of barium titanate near T_c (as predicted by Eq. 40), and observed experimentally by Samara) implies the strong variation of one or more zone center (q=0) optic-mode frequencies. The mode frequency and the dielectric constant are connected by Lyddane–Sachs–Teller (LST) relation

$$\varepsilon/\varepsilon_\infty = \Pi_i \ (\omega L)^2_i / (\omega T)^2_i \qquad (41)$$

where $\varepsilon\infty$ is the high frequency dielectric constant, and ωL and ωT are the longitudinal optic (LO) and transverse optic (TO) phonon frequencies for the various modes over which the product operation is carried out. It has been shown by Shirane (1970), that in many perovskite ferroelectric crystals nearly all of the temperature dependence of the (LO) and (TO) phonon frequencies arise from a single (TO) phonon, the ferroelectric mode. The ferroelectric mode in ABO$_3$ perovskites corresponds to the vibration of the positively charged B cation against the negatively charged oxygen octahedron.

Assuming a single (TO) phonon that varies with pressure, Samara (1970) showed that Eqs. (40) and (41) could be written as

$$\omega^2_f = \overset{*}{K}(p - p_0) \qquad (42)$$

where $\overset{*}{K}$ and p_0 are constants. An argument which was given by Samara to account for the increased "hardening" of ω_f with pressure will be given.

217

It is plausible to expect the restoring force and thus the frequency for this vibration mode to increase with decreasing interatomic distances. In other words, ω^2_f can be expressed as $\omega^2_{f^\infty}$ (short range forces – Coulomb forces). Since the short range forces are expected to be much more strongly dependent on interatomic distances than the long range Coulomb forces, the short range term increases faster than the Coulomb term, thus leading to an increase in ω_f. The pressure variation of the soft optic mode frequency in barium titanate is supported by direct measurement of the coupled accoustic–optic modes using combined Raman–Brillouin scattering (Peercy et al. 1973).

The piezoresistive effect in polycrystalline semiconducting ferroelectric barium titanate has been attributed qualitatively to the stress dependence of the grain boundary dielectric constant (Amin, 1986). A phenomenological analysis of the effect of internal stresses on the dielectric properties of ferroelectric polycrystalline barium titanate has been given by Buessem et al. (1966). A stress model has been proposed whereby each crystallite (grain) of barium titanate will be under an internal stress system of uniform compression along the tetragonal polar axis (c–axis) and under uniform tension along the two a–axes. Such a stress system is represented as follows

$$X_1 = X_2 = -X; \quad X_3 = X; \quad X_4 = X_5 = X_6 = 0$$

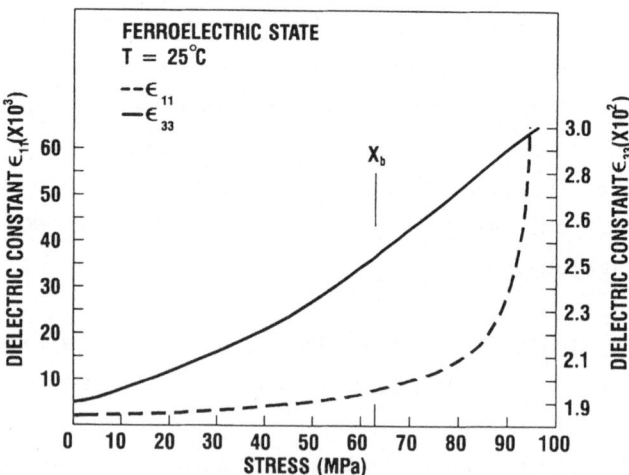

Fig. 6. Variation of ε_{11} and ε_{33} of $BaTiO_3$ with the stress system defined by Buessem et al. (1966).

The results of such a stress system on the dielectric properties of barium titanate as computed by Buessem et al. are displayed in Fig. 6. Based on this model, and assuming a dielectric constant of 3000 for fine grain barium titanate which corresponds to a stress level of 62 MPa, the calculated reduced lattice strain was found to be 15% in good agreement with x-ray results.

Janega (1986) has extended Buessem et al. treatment to account for the hydrostatic pressure effect on resistivity of semiconducting barium strontium titanate. The dielectric constant ε_{11} change due to applied stress was correlated to the resistivity changes via Eq. (6). The calculations were carried out using $BaTiO_3$ free energy coefficients and the phenomenological expressions given earlier.

In the ferroelectric state, the dielectric constant ε_{11} (along the tetragonal a-axis) was considered more significant (because of its large magnitude) than ε_{33} (along the polar axis) for barrier sensitivity to stress. Phenomenological computations of ε_{11} were carried as a function of temperature at different applied hydrostatic pressure levels, while maintaining the internal stress pattern of Buessem et al. (1966). The stress pattern at the grain boundary is expressed as follows

$$X_1 = X_2 = -X_b+p; \quad X_3 = X_b+p \tag{43}$$

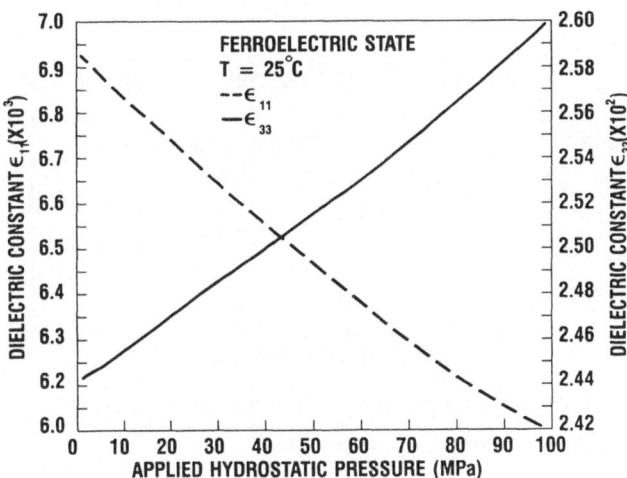

Fig. 7. Variation of ε_{11} and ε_{33} of $BaTiO_3$ with the stress system defined by Eq. (43).

where X_b is the internal stress level (approximately 62 MPa) in polycrystalline barium titanate as determined by Buessem et al. (1966), and p the applied hydrostatic pressure. Phenomenological calculations of the manner in which the ferroelectric state dielectric constants ε_{11} and ε_{33} vary with the stress pattern represented by Eq. (43) are depicted in Fig. 7.

In the paraelectric cubic state (above T_c), the spontaneous polarization, and the domain structure vanish, forcing the crystallites back to the stress free state [X_b=0 in Eq. (43)]. Figure 8 shows the hydrostatic pressure dependence of the paraelectric state dielectric constant at two temperature levels.

COMPARISON TO SILICON AND GERMANIUM

The piezoresistive effect under hydrostatic pressure conditions has been studied in silicon (Paul et al., 1955) and germanium (Paul et al., 1950). In germanium, the resistivity increases as the hydrostatic pressure is increased. This increase in resistance is explained as being due to both a decrease in mobility and to an increase in energy gap. For

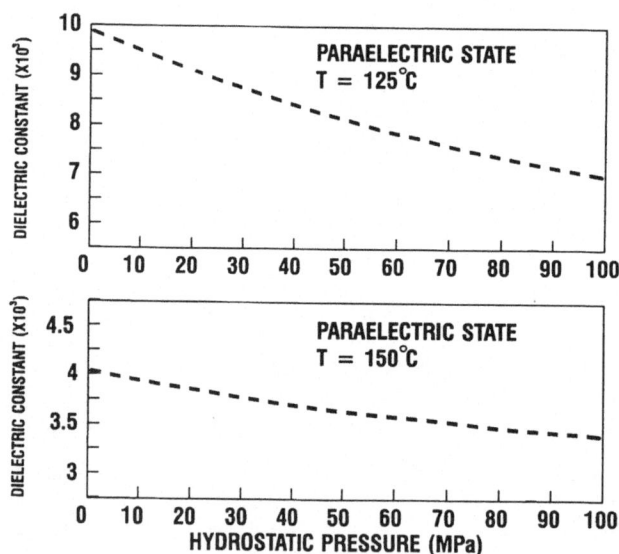

Fig. 8. Hydrostatic pressure dependence of BaTiOa$_3$-paraelectric state dielectric constant.

n-type germanium (δ = 35 ohm-cm) at room temperature, the mobility falls from 3900 cm^2/V-sec for small values of pressure to 900 cm^2/V-sec at 3 x 10^4 kg/cm^2. The energy gap (ΔE) increases at the rate 4.6 x 10^{-6} eV-cm^2/kg for pressures up to 1.5 x10^4 kg/cm^2, and less rapidly after that.

Typical values of the piezoresistive coefficient (x10^{-11} m^2/N) for silicon and germanium (Smith, 1954; Sauer et al., 1959) are, for n-type silicon (ρ=11.7 ohm-cm) are Π_{11} = -102, Π_{12} = + 54, Π_h = +6, and for p-type (ρ = 7.8 ohm-cm) Π_{11} = +90, Π_{12} = -42, Π_h = +6, for n type germanium (ρ=9.9 ohm-cm) Π_{11} = -90, Π_{12} = 41.5, Π_h =-7, and for p-type germanium (ρ = 15 ohm-cm) Π_{11} = 65, Π_{12} = -31, Π_h=+3. The hydrostatic piezoresistive coefficient Π_h is a linear combination of longitudinal and transverse coefficients (= Π_{11} +2 Π_{12}). The magnitude of these coefficient depends upon the crystallographic direction (slice orientation). The most sensitive direction is the <100> for n-type silicon, and the <110> for n-type germanium.

A qualitative argument for the microscopic origin of this behavior will be given. A full analytic account based on a non-degenerate band model is given by Herring (1955) and Herring et al. (1956) as due to strain induced carrier repopulation. On the basis of this model, the piezoresistive effect is attributed to inter-valley lattice scattering, i.e. to processes whereby a charge carrier is scattered from the neighborhood of one of the band edge points to the neighborhood of a different one.

For example, in n-type silicon, a uniaxial stress (X_1) parallel to a valley will have the effect of raising the energy minima along the direction of the stress and lowers the four valleys in a direction perpendicular to the stress. Since the four perpendicular valleys have lower energy, electrons tend to flow from the higher energy levels to the lower levels, and the relative population is given by Maxwell-Boltzmann's statistics. Since the perpendicular valleys have the larger electron population, a field applied in the direction of stress will result in a higher electron mobility. This will increase the conductivity and decrease the resistivity in proportion to the applied stress as observed experimentally.

In p-type silicon, the resistivity increases with applied stress, also, as doping increases the holes get closer, and the type of statistics changes from Maxwell-Boltzmann to Fermi-Dirac. As a consequence, the piezoresistive coefficient no longer varies inversely with temperature, but less rapidly. For a doping level of 10^{20} at/cc the piezoresistive coefficient decreases and becomes less temperature sensitive (+/- 3% over the temperature range from 0 to approximately 200°C).

Piezoresistance in both n- and p-type polycrystalline silicon has been considered by French et al. (1985). On the basis of their model, both bulk and grain boundary (Schottky-type) are contributing to the observed piezoresistance. Grain size, orientation, trap density, and doping level are among the important parameters that affect the piezoresistance coefficient.

In semiconducting polycrystalline barium titanate, grain boundary barriers are temperature, voltage, frequency, and stress sensitive. In the paraelectric state, the piezoresistive effect could be explained "semiquantitatively" in terms of a simple barrier model and the stress dependence of the barrier layer dielectric constant (Janega, 1986). This model does not take into account any bulk (grain) contribution to the piezoresistive effect. In other words, it assumes that piezoresistivity is wholly a barrier layer effect.

In the ferroelectric state, however, the situation is exceedingly complex due to the presence of ferroelectric domains, and the uncertain nature of the elastic boundary conditions. It has been shown that both ferroelectric domains and mechanical stresses which exist at the grain boundary layer contribute significantly to barrier height. Unfortunately, in polycrystalline ferroelectrics the exact pattern of both is not known. In addition, the behavior of domain walls, and phase boundary motion at high pressure is a very complex phenomenon.

Additional experimental and theoretical work will be required to answer the above questions, delineate the precise nature of the interfacial acceptor states, and analyze the influence of grain size and dopant concentration on the piezoresistive properties of semiconducting barium titanate.

REFERENCES

Aizu, K., 1972, Electrical, mechanical, and electromechanical orders of state shifts in nonmagnetic ferroic crystals, <u>J. Phys. Soc. Jpn.</u>, 32:1287.

Aizu, K., 1973, Second order ferroic state shift, <u>J. Phys. Soc. Jpn.</u>, 34: 121.

Amin, A., and Newnham, R. E., 1980, Tertiary ferroics, <u>Phys. Stat. Sol(a).</u>, 61:215.

Amin, A., and Shukla, V., 1985, Effects of mechanical processing on semiconducting properties of barium titanate, <u>J. Am. Ceram. Soc.</u>, 68(7):C-167.

Amin, A., Spears, M., and Kulwicki, B. M., 1983, Reaction of anatase and rutile with barium carbonate, <u>J. Am. Ceram. Sos.</u>, 66(10):733.

Amin, A., 1986, Piezoresistivity in semiconducting perovskites, <u>TI Engr. J.</u>, 3(2):38.

Amin, A., 1987, Computer-controlled system for investigating the hydrostatic piezoresistive effect as a function of temperature, <u>Rev. Sci. Instrum.</u>, 58(8):1514.

Amin, A., and Cross, L. E., 1985, Effect of electric boundary conditions on morphotropic $Pb(Zr,Ti)O_3$ piezoelectrics, <u>Jpn. J. Appl. Phys.</u>, Suppl. 24-2: 229.

Amin, A., Newnham, R.E., and Cross, L.E., 1986, Effect of elastic boundary conditions on morphotropic $Pb(Zr,Ti)O_3$ piezoelectrics, <u>Phys. Rev. B.</u>, 34(3):1595.

Bhagavantam, S., 1966, "Crystal Symmetry and Physical Properties," Academic Press, New York.

Buessem, W. R., Cross, L. E., and Goswami, A. K., 1966, Phenomenological theory of high permittivity in fine-grained barium titanate, <u>J. Am. Ceram. Soc.</u>, 49(1):36.

Buessem, W.R., Cross, L.E., and Goswami, A. K., 1966, Effect of two-dimensional pressure on the permittivity of fine-and coarse-grained barium titanate, <u>J. Am. Ceram. Soc.</u>, 49 (1): 36.

Chan, H. M., Harmer, P. M., and Smyth, M. D., 1986, Compensating defects in highly donor-doped $BaTiO_3$. <u>J. Am. Ceram. Soc.</u>, 69 (6): 507.

Cochran, W., 1960, Crystal stability and theory of ferroelectricity, <u>Adv. Phys.</u>, 9: 387.

Daniels, K., Haerdtl, H., and Wernicke, R., (1978/1979), The PTC effect of barium titanate, <u>Philips Tec. Rev.</u>, 38:73.

Devonshire, F.A., 1949, Theory of barium titanate-part I, <u>Phil Mag.</u>, 40:1040.

Devonshire, F. A., 1951, Theory of barium titanate–part II. Phil. Mag., 42:1065.

French, P. J. and Evans, A. G. R., 1985, Polycrystalline silicon strain sensors, Sensors and Actuators, 8: 219.

Goodman, G., 1963, Electrical conduction anomaly in samarium–doped barium titanate, J. Am. Ceram. Soc., 46 (1): 48.

Goswami, A. K., and Cross, L., E., 1968, On the pressure and temperature dependence of the dielectric properties of perovskite barium titanate, Phys. Rev., 171(2):549.

Henisch, K. H., 1984, "Semiconductor Contacts," Oxford.

Herring, C., 1955, Transport properties of many–valley semiconductor, Bell System Tech. J., xxxiv(2): 237.

Herring, C., and Vogt, E., 1956, Transport and deformation–potential theory for many–valley semiconductors with anisotropic scattering, Phys. Rev., 101(3): 944.

Heynecek, J., 1974, Elastoresistance of n–type silicon on sapphire, J. Appl. Phys., 54(6): 2631.

Heywang, W., 1961, Barium titanate as a semiconductor with blocking layers, Solid–State Electron., 3 (1): 51.

Heywang, W., 1963, Behavior of reactance of $BaTiO_3$–cold conductors as a confirmation of the model with blocking layers, Z. Angew. Phys., 16 (1): 1.

Heywang, W., 1964, Resistivity anomaly in doped barium titanate, J. Am. Ceram. Soc., 47 (10): 484.

Heywang, W., and Gunterdorfer, M., 1966, On the resistivity of doped BaTiO3, Proc. Intern. Meeting Ferroelectricity, 2: 307.

Heywang, W., 1971, Semiconducting barium titanate, J. Mat. Sci., 6: 1214.

Igarashi, H., Michiue, M., and Okazaki, K., 1985, Jpn. J. Appl. Phys., suppl. 24–2: 305.

Janega, L. P., 1986, Hypothesis to explain pressure effects on resistivity in semiconductive barium titanate ceramics, Solid–State Electron., 29 (1): 59.

Jonker, H. G., 1964, Some aspects of semiconducting barium titanate, Solid–State Electron., 7: 895.

Jonker, H. G., 1967, Halogen treatment of barium titanate semiconductors, Mat. Res. Bull., 2: 401.

Khan, M., 1971, Effect of heat treatment on the PTCR anomaly in semi–conducting barium titanate, Am. Ceram. Soc. Bull., 50 (B): 676.

Kulwicki, M. B., and Purdes, J. A., 1970, Diffusion potentials in barium titanate and the theory of PTC materials, Ferroelectrics, 1:253.

Lewis, V. G., Catlow, A. R. C., and Casselton, W., E., R., 1985, PTCR
        effect in Ba TiO$_3$, J. Am. Ceram. Soc., 68 (10): 555.

Mallick, Jr., T. G., and Emtage, R. P., 1986, Current voltage
        characteristics of semiconducting barium tiatnate ceramic, J. Appl.
        Phys., 39 (7): 3088.

Mason, P. W., and Thurston, N. R., 1957, Use of piezoresistive materials
        in the measurement of displacement, force, and torque, J. Acoust.
        Soc. Am., 29 (10): 1096.

Nye, J. F., 1976," Physical Properties of Crystals and Their
        Representation by Tensors and Matrices," Cambridge, London.

Paul, W., and Pearson, L. G., 1955, Pressure dependence of the
        resistivity of silicon, Phys. Rev., 98: 1755.

Paul, W., and Brooks, H., 1954, Pressure dependence of the resistivity of
        germanium, Phys. Rev., 94: 1128.

Peercy, P. S., and Samara, G. A., 1973, Pressure and temperature
        dependences of the dielectric properties and Raman spectra of
        RbH$_2$PO$_4$., Phy. Rev., 88: 2033.

Saburi, O., 1960, Piezoresistivity in semiconductive barium titanate, J.
        Phys. Soc. Jpn., 15: 733.

Samara, G. A., 1966, Pressure and temperature dependences of the
        dielectric properties of the perovskites BaTiO$_3$ and SrTiO$_3$., Phys.
        Rev., 151(2): 378.

Samara, G. A., 1970, The effect of hydrostatic pressure on ferroelectric
        properties, J. Phys. Soc. Jpn., suppl. 28: 399.

Samara, G. A., Sakudo, T., Yoshimitsu, K., 1975, Important generalization
        concerning the role of competing forces in displacive phase
        transitions, Phys. Rev. Lett., 35 (26): 1767.

Sauer, A. H., Flaschen, S. S., and Hoesterey, C. D., 1959,
        Piezoresistance and piezocapacitance effect in barium strontium
        titanate ceramics, J. Am. Ceram. Soc., 42 (8): 363.

Scholl, E. J., 1986, Lowering of grain – boundary heights by grain
        curvature, J. Appl. Phys., 60 (4): 1434.

Shirane, G., 1970, Neutron inelastic scattering study of soft modes, J.
        Phys. Soc. Jpn., suppl. 28: 20.

Smith, S. C., 1954, Piezoresistance effect in germanium and silicon,
        Phys. Rev., 94 (1): 42.

Yasukawa, A., Shimada, S., Yoshitaka, M., and Kanda, Y., 1982, Design
        considerations for silicon circular diaphragm pressure sensors, Jpn.
        J. Appl. Phys., 21 (7): 1052.

Zaima, S., Yasuda, Y., Kawaguchi, S., Tsuneyoshi, M., Nakamura, T., and Yoshida, A., 1986, Piezoresistance in n-channel inversion layers of silicon films on sapphire, J. Appl. Phys., 60 (11): 3959.

DOMAIN PATTERNS IN HELICAL MAGNETS

Walter M. Fairbairn

Department of Phyiscs
University of Lancaster
Lancaster, U.K.   LA1 4YB

INTRODUCTION

When magnetic systems order it is well-known that on a macroscopic scale the magnetisation will normally not be uniform and parts of the sample will order differently, producing a spatial pattern in which individual regions can be associated with specific well-defined ordering. These ordered regions (the domains) co-exist with their adjoining boundaries, classified as walls, forming an identifiable pattern within the magnetic sample. There has been a considerable amount of experimental and theoretical investigation of these domains and their associated walls in ferromagnetic materials but the corresponding properties for other types of magnetic systems have not received such detailed consideration.

Ferromagnetic systems are relatively simple in that the order parameter within a single domain is defined by a magnitude which depends on the temperature but is the same over all of the sample and a direction which is associated with the domain and, because of the structure of the material, is likely to be restricted to only a few possibilities. On passing between domains the direction of the order parameter (the magnetization in the case of ferromagnetic materials) changes. The region over which this change takes place is a domain wall. The structure of these walls is well-defined. Types of wall which have been analysed most are those within which the direction changes by 180^0 and of which the two best known examples are the Bloch wall and the Néel wall. Their size (normally known as width) and their energy depend on the magnitude of physical quantities such as the exchange interactions the anisotropy and the magnetostriction of the material.

The samples of ferromagnetic materials are three-dimensional and the domain structure is usually investigated by examination of the topology of the magnetic structure on the surfaces of the sample. Techniques to do this use photons, electrons and neutrons to obtain images of the domains. In most specimens the dominant pattern for domains is that of bubbles or stripes. The configuration is likely to depend on defects and other pinning mechanisms but even in the purest of samples a domain pattern develops, because on ordering the nucleation process causes domains with their associated walls to form in what is strictly a metastable structure. With three-dimensional ferromagnetic materials the walls can have the most complicated geometry. It is true that a striped pattern occurs in most specimens under appropriate conditions but it is not generally the case that the domains have plane parallel faces; the walls frequently have curvature.

Magnetic materials which order in a helical phase may provide a convenient proving ground for testing predictions about domain walls. This is particularly true if the materials are of layered structure, either by specific manufacture or because of their crystalline structure. The latter occurs for the heavier rare earth metals and many of their alloys[1], with the hexagonal-close-packed lattice having a preferential direction (the c-axis) which thereby picks out within the structure a set of parallel planes normal to it. When these materials do order magnetically each of these planes seems to behave as a unit so that the system is quasi-one-dimensional with each plane normal to the c-axis forming a single magnetic entity. Any domain pattern which is formed has walls also normal to the c-axis. The behaviour of such geometrically well-defined walls is more easily predicted.

Although these magnetically-ordered systems are not ferromagnetic each of the constituent planes is ordered ferromagnetically. Successive planes (i.e. layers) have their direction of magnetization rotated relative to that of their neighbours. The systems have a helical ordering with the axis of the helix being the c-axis of the structure and the pitch of the helix being given by the angle between the magnetization of successive ferromagnetically ordered planes. This turn-angle is determined by the sizes and signs of the couplings between magnetic moments and these are of relatively long range; for rare earth metals this dominant interaction is RKKY, which can be effective out to fifth or sixth nearest neighbours[2].

The turn angle can be either positive or negative, resulting in helices which are either right-handed screw or left-handed. This chiral property means that there are only two types of domain possible in a specimen which is a single crystal. Also the walls between domains are geometrically simple, being composed of parallel planes of sites. Helical magnets provide real examples of the simplest model of a physical system which undergoes a second-order phase transition because only two types of domain are possible (right-handed and left-handed screw helices), the domain walls have planar boundaries and the three-dimensional material behaves to a good approximation in a one-dimensional manner.

THE DOMAIN WALLS

The appropriate physical system can be described by a one-dimensional chain of magnetic units (planes) which couple through the long-range effective RKKY exchange interactions $J(n)$. In accordance with experimental evidence there exists also an effective anisotropic interaction A which encourages the magnetic moments to align perpendicular to the length of the chain. If an applied external magnetic field is assumed to be present then the Hamiltonian for such a system can be written as

$$H = \sum_{all\ k} \left[\sum_{n \geqslant 1} J(n)\ \underline{S}_k \cdot \underline{S}_{k+n} + A\ S_{zk}^2 + g\mu_b\ \underline{S}_k \cdot \underline{B} \right]$$

g being the appropriate g-factor with applied field $\underline{B}$, the z-axis being the axis of the chain and k labelling the magnetic units. This Hamiltonian will determine the structure of the domains and of the walls.

First we consider the system with no external field ($B \equiv 0$). The ordered state of the system will be a helical structure with turn-angle γ determined as a non-zero solution of[2]

$$\sum_n nJ(n)\ \sin n\gamma = 0 \tag{1}$$

There are two solutions $+\gamma$ and $-\gamma$ which define the two possible domains within the system. If the interplanar distance is a, then the magnetic moments in the ordered state all align perpendicular to the z-axis ($S_{zk} = 0$) and in the plane k at an angle $\phi = k\gamma$ (or $- k\gamma$) to some fixed arbitrary direction within the easy plane. More formally $d\phi/dz = \gamma/a$ (or $- \gamma/a$) within a domain.

For the wall the simplest assumption is that within that region also the magnetic moments remain aligned perpendicularly to the z-axis and the structure of the wall is determined by how ϕ varies. Clearly only the relative values of the angle ϕ are important and the relevant equations are most sensibly written in terms of $u = d\phi/dz$. In the continuum approximation with $a = 1$ the appropriate ordinary differential equation for the static configuration obtained from the more general equations of motion when derivatives higher than the third are assumed to be negligible[3], is

$$\left[\sum_n n^4 J(n) \cos nu\right] \frac{d^3u}{dz^3} - 2 \left[\sum_n n^5 J(n) \sin nu\right] \frac{d^2u}{dz^2} \frac{du}{dz}$$

$$- \frac{1}{2} \left[\sum_n n^6 J(n) \cos nu\right] \left(\frac{du}{dz}\right)^3 + 12 \left[\sum_n n^2 J(n) \cos mu\right] \frac{du}{dz} = 0 \qquad (2)$$

As expected this is a highly non-linear equation. The structure of a domain wall is obtained from the solution which obeys the boundary conditions: $u = - \gamma$ and all derivatives of u zero at large negative z, $u = +\gamma$ and all derivatives of u zero at large postive z. Then equation (2) can be integrated twice to obtain a first-order equation, again highly non-linear,

$$\left[\sum_n n^4 J(n) \cos nu\right] \left(\frac{du}{dz}\right)^2 = 24 \left[\sum_n J(n) (\cos nu - \cos n\gamma)\right] \qquad (3)$$

with the boundary conditions $u = \gamma$ at large z and $u = 0$ at $z = 0$, because the solution $u(z)$ must be an odd function of z. This equation normally needs to be integrated numerically. A typical solution is shown in Figure 1.

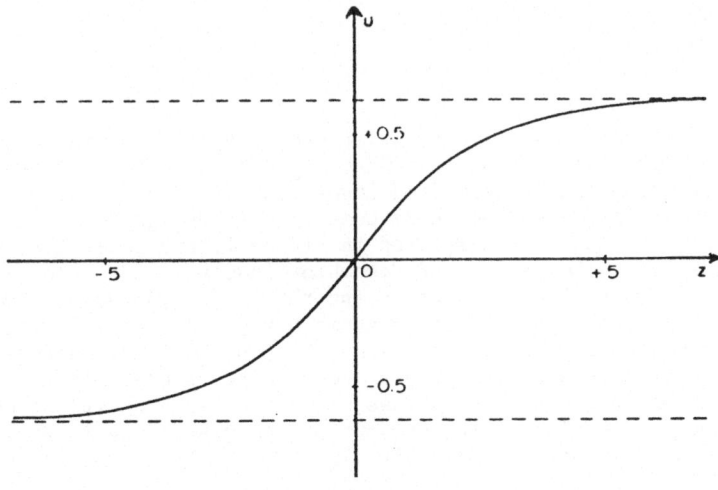

Fig. 1. Variation with position of the turn-angle u within a domain wall between two domains of opposite chirality. The horizontal axis z denotes distance in units of lattice-plane separation and the curve shown corresponds to the rare earth metal Ho at 50K. The centre of the wall ($z = 0$) in this case occurs midway between two lattice planes.

229

For helical magnets it is clear that u, the gradient of the direction of magnetization, is the appropriate order parameter and that the structure of a domain wall is defined by how that quantity varies through the wall. The above description is based on the assumption that the magnetic moments are always restricted to lie perpendicular to the wall; in a sense this type of wall is the analogue of Bloch walls in ferromagnets. However even although the ordered domain contain only moments which are confined in this way it may be that such a restriction is inappropriate for the walls : compare the Néel walls in ferromagnets. When the moments are permitted to deviate from directions within the easy plane this new degree of freedom can result in structures which are more complicated than that described by (3) and Figure 1.

STATIC APPLIED FIELDS

From energy considerations the static wall in no applied field is that discussed above and defined by equation (2) for $u(z)$. The application of a static uniform magnetic field has little effect. A field within the easy plane will modulate the turn angle within domains and walls[4] but is unlikely to have much effect on the pattern of domains, although a large enough field will produce a ferromagnetic ordering throughout the whole sample[5]. When the field is in the hard direction a small canting (of all moments) will occur but again the domain pattern and the turn-angles are not affected, although the width of the walls is increased slightly[5]. The domains and associated walls are very stable in the presence of applied fields which are both static and uniform. This is not true when the applied fields vary with time and the effect of time-dependent magnetic fields is reported in the next section.

Many analytic discussions of domain walls are based on the soliton solutions of non-linear differential equations, with the sine-Gordon and ϕ^4 equations being characteristic examples. The first integral of equation (2) is

$$2 \left[\sum_n n^4 \; J(n) \; \cos nu \right] \frac{d^2u}{dz^2} - \left[\sum_n n^5 \; J(n) \; \sin nu \right] \left[\frac{du}{dz} \right]^2$$

$$+ \; 24 \left[\sum_n n \; J(n) \; \sin nu \right] = 0 \tag{4}$$

which can be compared, as a second-order ordinary differential equation, with those for the time-independent sine-Gordon and ϕ^4 theories. As for them, a further integration can be performed (equation (3)), but no general analytic solutions are available so that only numerical investigations are possible. A solution so obtained, and displayed in Figure 1, indicates that soliton-type solutions of (4) do exist and resemble closely in form the tanh-type solutions of the sine-Gordon equation and the $\tan^{-1}$-type solutions of the ϕ^4 equation. It is known[6] that these systems have localized solition-type solutions some of which involve the excitation of internal degrees of freedom. For the sine-Gordon systems these are known as breathers. There is no clear indication of whether equation (4) has solutions of this type or not. However when the moments are allowed to move out of the easy plane it will be found that similar excitations are possible.

TIME-VARYING APPLIED FIELDS

The most-straightforward case to consider is that of an applied magnetic field along the direction of the hard axis, which is the c-axis for the rare earth metals and their alloys. Such a field will produce a canting of the magnetic moments out of the easy plane, the magnitude depending on the strength of the field. It will cause also the moments to rotate within the easy plane; $\dot{\phi}$ will not be zero. However all moments will be equally affected so that the domain pattern will not change.

When this applied field changes with time both the rate of rotation ($\dot{\phi}$) and the canting angle (ε) will alter but the only effect on the walls will be an extremely small change to their width. Nevertheless within the walls, but not within the domains, there is a dependence of the value of $u = \partial\phi/\partial z$ on the canting angle so that changes to ε can affect the internal structure of the wall through variation in $u(z)$. Perhaps the most significant of such effects is that an oscillating applied magnetic field at appropriate frequencies, which are likely to be in the GHz range, should excite internal modes of vibration within the wall[5]. These are resonant oscillations of the order parameter u which are confined to the domain walls. The overall pattern of the domain structure would not be affected but energy would be absorbed from the time-varying field to excite a cooperative oscillation within the walls. Although the width of the wall could change slightly there would be no overall displacement induced by such fields. The excitation involves the coupled oscillations of canting angle ε and turn angle u.

TEMPERATURE VARIATION

It is well-established that the pitch (turn angle) of the helix in the ordered state changes with temperature. In most materials on cooling through an upper ordering temperature the helical phase is obtained with a specific turn-angle. On further cooling, this angle γ decreases until at a lower temperature there is a second transition, this time into a ferromagnetic phase[1]. From equation (1) the change in γ is driven by any changes in the exchange parameters $J(n)$ which describe the long-range interactions within the material. It should be noted that γ has not normally attained the value zero when the transition into the ferromagnetic phase occurs.

Because the detailed geometry of the wall is determined, through equation (2), also by the values of the exchange parameters $J(n)$ the width should vary with temperature. The variation in γ is not large in most cases and there is unlikely to be enormous changes in any of the $J(n)$. However because in some of the formula the more distant (larger n) interactions are quite heavily weighted, by factors of n^4 for example in (3), slight changes in what are usually small quantities could have an appreciable effect.

Nevertheless it does seem to be unlikely that variation in the values of the exchange parameters as the temperature is lowered will cause the domain walls in the helically ordered phase to expand sufficiently to drive the system into the ferromagnetic phase as has been suggested[7,8]. There is also no experimental evidence to support such a dramatic change, unless it occurs very close to the lower transition temperature.

NON-UNIFORM APPLIED FIELDS

Whereas uniform applied magnetic fields, either static or time-varying, along the hard-axis direction or small applied fields within the easy-plane do not affect the pattern of domains a non-uniform field will. For simplicity consider a static field in the z(hard-axis) direction. This will produce a different dynamic effect at different sites within both the domains and the walls, because ϕ will be changing, with time, at different rates. It follows that $u = \partial\phi/\partial z$ will not longer be uniform throughout what was previously considered to be a single domain.

If it is realistic to assume that all magnetic moments continue to lie within the easy-plane then the domain structure still may not be changed radically because with $\varepsilon \equiv 0$ the equations of motion reduce to (2) together with

$$\frac{\partial u}{\partial t} = - g \, \mu_b \left(\frac{\partial B_z}{\partial z} \right)$$

It follows that if the field gradient is uniform the order parameter u is affected similarly at all parts of the sample. Each domain will remain a region in which the order parameter has a uniform value, although that value will be changing with time. In some domains the turn angle will be increasing in others it will be decreasing. Some helices will be wound up while those of opposite chirality will be unwound. Domain walls also will be unwound and the specimen will eventually be helically magnetised with the same chirality overall, but not uniformly.

For magnetic fields applied along the hard axis direction it is only those with non-uniform field gradient which will change the pattern of the helical domains.

DISCUSSION

The pattern of domains in helical magnetic materials is expected to be both simple and stable. Because many of those materials have a structure with one principal axis, such as hcp, the domains order with the axes of the helices along the direction of this principal structural axis. Only two possible types of domain can then exist, one being a right-handed screw, the other a left-handed, but both having the same turn-angle which will, however, vary with temperature. The structure of the associated domain walls is well-determined, by equation (2), and for most materials for which estimates have been made the width of the walls is not great, being less than twenty lattice planes. This pattern of domain structure has been seen using spin-polarized neutron topography in rare earth metals and alloys[4,9], and some other materials such as MnP.

These helical phases exist over a range of temperatures, frequently associated with a lower temperature ferromagnetic phase. When samples are temperature-cycled between these phases, and also into and through the disordered paramagnetic phase, the geometric pattern of the domains is frequently reproducible although this is more pronounced at the ferro-magnetic-helical phase boundary than at the boundary between the paramagnetic and helical phases. Any such reproducibility is likely to be governed by impurities and other defects in the sample.

What is as intriguing is the suggested stability of the domain pattern in respect of applied magnetic field. It is not that the domains and walls are unaffected by such external fields but it seems that any effects occur in a manner which retains the identifiable characteristics of the domains. This is more true for fields applied along the hard axis than for those within the easy plane. The latter produce destabilizing effects when the fields are large enough. Clearly a strong field in the easy plane will produce preferential alignment in that direction and the helical domains will be destroyed[5].

Another intriguing aspect relating to these domain walls is the possibility of excitation of their internal degrees of freedom by oscillating applied fields. For helical walls such excitations involve collective oscillations of the magnetic moments out of the easy plane. The helical structure relative to the hard axis direction is little changed, so that this effect is not identical with the transitions suggesgted[10] as possible within the domain walls of ferromagnetic materials. In this latter case there is a change from one type of wall to another. For helical magnets only one form of domain wall is realistically allowed and the field-induced changes are most readily thought of as internal excitations.

It would appear that helically-ordered magnetic materials provide the prospect of very stable domain structures, in respect of both applied field and temperature variations. Once the specimen has ordered it is not easy to change the basic topology of the pattern of domains. They may mean that such materials are relatively uninteresting because for physicists it is a changing system that is more exciting. However it is possible to alter quite radically the internal structures of the domains and walls using both non-uniform and time-varying fields. Observation of these latter transitions would be both fascinating and might lead to applications of these materials. The coupling between spatial and magnetic structures could cause the production of very high frequency phonons when the internal modes within the domain walls are excited.

Whether the recently discovered[11] magnetic superlattices containing rare earth metals with their magnetic layers and ordering interactions diluted by intervening non-magnetic layers possess similarly stable domain patterns remains to be investigated.

This article is dedicated to Heinz Henisch on the occasion of his sixty-fifth birthday, with gratitude for his continual insistence that all worth-while mathematical and computational analyses should be applicable to real physical systems.

REFERENCES

1. W.C. Koehler, Magnetic Structures of Rare Earth Metals and Alloys, in "Magnetic Properties of Rare Earth Metals", R.J. Elliott, ed., Plenum, New York (1972).
2. S. Singh, "Magnetic Structure of Holmium", Thesis, University of Lancaster (1985).
3. W.M. Fairbairn and S. Singh, "Domain Walls in Helical Magnets", J. Mag. and Mag. Mat. 54-57 : 867 (1986).
4. A. Drillat, "Recherches d'Effets de Commensurabilite et Domaines de Chiralite dans la Phase Helimagnetique", Thesis, University of Grenoble (1983).
5. W.M. Fairbairn, "Motion of Domain walls in Helical Magnets", J. Appl. Phys. 61 : 4123 (1987).
6. J.F. Currie, J.A. Krumhansl, A.R. Bishop and S.E. Trullinger, "Statistical Mechanics of One-dimensional Solitary-wave-bearing Scalar Fields", Phys. Rev. B22 : 477 (1980).
7. A. Del Moral and E.W. lee, "The Reversible Susceptibilities of Dysprosium and Terbium", J. Phys. F4 : 280 (1974).
8. S.B. Palmer, "Antiferromagnetic Domains in Rare Earth Metals and Alloys" J. Phys. F5 : 2370 (1975).
9. J. Baruchel, S.B. Palmer and M. Schlenker, "Chirality Domains in Helical Antiferromagnets", J. de Physique 42 : 1279 (1981).
10. J. Lajzerowicz and J.J. Niez, J. de Phys. Lettres 40 : 165 (1979).
11. J.J. Rhyne, R.W. Erwin, J. Borchers, S. Sinha, M.B. Salamon, R. Du and C.P. Flynn, "Occurrence of Long-range Helical Spin Ordering in Dy-Y layers", J. Appl. Phys. 61 : 4043 (1987).

SILICON NITRIDE FILMS FORMED WITH

DC-MAGNETRON REACTIVE SPUTTERING

Napo Formigoni

Energy Conversion Devices
Troy, Michigan

ABSTRACT

A dc-magnetron reactive sputtering process, that employs elemental Si and N, has been developed for the deposition of silicon-nitride films.

A Balzers vacuum plant model BAS450PM, equipped with cryopump and dc-magnetron has been adapted for this task. In order of importance the significant features of this process are:

a)   The substrate temperature during deposition is approximately 50 deg. C
b)   The inlet for Ar is located near the Si target and the one for N2 near the substrate
c)   The substrates are revolving during deposition, in order to obtain controlled and uniform film properties
d)   The film deposition rate is 1 angstrom per revolution, or about 30 angstrom/min.
e)   Several 3-in. dia. round (or otherwise different substrates) can be simultaneously coated.

This film exhibits excellent mechanical strength, high corrosion resistance, low electrical leakage and high dielectric breakdown. Therefore, it has been successfully employed for the encapsulation and passivation of photosensitive amorphous semiconductor films, such as optical memories (1) and electron devices such as thin film OTS switches (2).

The silicon nitride by this process is not perfectly stoichiometric, since Auger analysis has found it nitrogen deficient. Its refractive index by ellipsometry is 1.85-to-1.87, while the refractive index of bulk Si3N4 is 2.00. The very low deposition temperature, which is probably responsible for the incomplete chemical reaction of Si and N2.

INTRODUCTION

Practically every electron device in existence needs a protective coating in order to operate properly. Without such protection, atmospheric humidity causes corrosion and ultimately permanent device damage. Amorphous electron devices are no exception, therefore a protective layer must be placed on their surface. Moreover, amorphous glasses like those

TABLE I.

PREPARATION METHODS FOR ELECTRONIC
AND OPTICAL DIELECTRIC THIN FILMS

MATERIALS	PROCESS	SUBSTRATE TEMPERATURE	CORROSION	MATERIAL COST
SiO_2, TiO_2, TiN	Thermal	High	High	Low
Ta_2O_5	Anodization	Low	Medium	Low
Si_3N_4, SiO_2	Plasma-CVD	Medium	High	High
Si_3N_4, SiO_2	E-Beam	High	Low	Medium
SiO	Thermal Evaporation	Medium	Low	Medium
Si_3N_4, SiO_2, Al_2O_3	RF-Sputtering	Medium	Low	High
$Si_3N_4, SiO_2 Al_2O_3$	DC-Magnetron Reactive Sputtering	Low	Low	Low

employed in OTS thin film devices contain elements like S and As which at the same time are highly reactive in the chemical sense and have a low melting point. Therefore the choice of a suitable passivating film, and of its formation process is not a trivial one.

Table I lists a number of known methods for depositing thin film dielectrics, that are used in the electronic industry for device passivation. Among them dc-magnetron reactive sputtering appears to be the best method for such a film, because the deposition temperature is low, the process gases will not (at that temperature) alter the device and the material cost is modest. Actually a large polycrystalline, high purity silicon target costs several thousand dollars but lasts a long time.

EXPERIMENTAL

The Balzers model BAS450PM, is the vacuum plant used for depositing by reactive sputtering the silicon nitride film. Its cylindrical vacuum chamber, 45 cm dia. and about 50 cm high, is designed for side sputtering and has four ports for targets that can be sequentially operated, without breaking vacuum. The chamber design offers an ingenious combination of target compartments and coaxial rotating shutter. This permits doing individual getter sputtering on each target, which can bury target's surface contaminants, without cross depositing on the other targets or the substrates. This chamber is designed for uniformly coating up to 24 three-in. dia. silicon wafers mounted on a cage which rotates with uniform angular velocity in front of the operating target. However, this part was modified to suit our process requirements, as will be described later.

Pumping is done by the combination of two cryogenic stages: the 77Kelvin LN2 and the 14Kelvin He cryosorption stage. Initial concern about "poisoning" of the cryosorption panels by chalcogenide evaporants, like S, Se, Te or by As have simply not materialized in more than two years of daily operation, regular weekly regeneration, and occasional maintenance.

The target is of the indirect cooling type, where the cooling circuit is entirely outside the vacuum chamber. Its size is 5" x 10" and the sputtering surface is designed to operate about 1/2 inch in front of the cooling plate (3/4 inch in front of the magnets).

The first several months of this work were spent in developing the technique for depositing dielectrics, using dc-magnetron reactive sputtering. Pertinent information in the literature (3), (4), (5) was still scant two years ago, and what little was available rarely went beyond the statement of results and broad general guidelines on the technique. The passivation of the target, under the action of the reacting gas was a primary concern. But this did not happen to a significant extent, probably because under the intense Ar bombardment, afforded by this magnetron, the chemical compounds formed on the target are continuously decomposed. For example, an aluminum oxide can be deposited with our Balzers plant by reactively sputtering an Al target in pure O_2, without any Ar if the target power is sufficiently high. Likewise a graphite target can easily be sputtered in pure O_2. Negative O_2 ions however will heavily bombard the substrates and substantially raise their temperature during deposition. In conclusion, experience has shown that the Balzers dc-magnetron can be operated very successfully in most conditions. The only notable exceptions are the targets of insulators or near insulators, such as low conductivity chalcogenide alloys known also as OTS glasses. In such instances, rf-sputtering must be used.

Careful observation of the effect of O2 and N2 on the dc-magnetron discharge, as well as on the properties of the deposited films was important in determining the configuration of the injectors. Argon is injected close to each target, but a rather complex injecting apparatus (shown later on) had to be designed for uniform gas distribution and best ionization effectiveness. The reactive gases at the contrary should be released close to the deposition substrate so as fully to react with the deposited material, before their concentration drops to equilibrium values. In our system the preferred gas operating pressures are found in the range 0.1-1.0 Pa, where the mean free path for Ar is about one centimeter.

In dc and rf diode sputtering the reactive gases are generally injected premixed with Ar at 1-to-10% concentrations. Instead with dc-magnetron, only the injection of straight N2 and O2 proved to be effective for obtaining thin dielectric films without appreciable optical absorption in the visible. Probably the low substrate temperature, characteristic of magnetron operation, slows down the chemical reactions and for this reason the highest possible reacting gas concentration is required as a remedy.

Four materials were investigated as passivating dielectrics candidates. The results of this study, summarized in table II, points to the conclusion that silicon-nitride among the four dielectrics listed in the table offers the best combination of easy processing and optimal dielectric properties.

THE REACTIVE SPUTTERING APPARATUS

The conceptual layout for the planar dc-magnetron reactive sputtering apparatus, adapted from the original Balzers, is shown in figure 1. From left to right one can see the planar magnetron with the injector for Ar fastened to the target dark space shield, the cylindrical shutter, the mask and the substrate carrier disc. An electric motor outside the vacuum chamber actuates the driving shaft through a rotary vacuum seal. A pair of bevel gears transmits the rotation to a circular disc on which the substrates are fastened with springs. The mask has a triangular window, shown in figure 2, which is the plan view of the apparatus. This mask is interposed between the planar magnetron target and the substrates, which are located 1/8 in. behind the mask and revolve at several tens rpm, as mentioned before. The sputtered material can only deposit on the substrate through the mask window, which spans a 40 degrees sector of the disc. The thickness of the deposited film on the substrates is uniform if the material flux from the target is uniform over the window. The distance of nearly three inches between target surface and revolving substrates helps to smooth out the uneven emission from the magnetron target. With this configuration all areas of the disc are exposed to the steady and uniform flux of atoms for the same length of time during each revolution. Near the mask boundaries however the flux from the target is smaller, because of the smaller target viewing angle from these areas. Therefore, the deposition uniformity will fall off in the extreme boundary areas of the mask. Since the deposition time is only approximately 1/10 the revolution time, the deposited material can come to equilibrium with the substrate before the next shower of deposit arrives. With a deposition rate of, for example, 15 angstrom/min and the substrate revolving at 60 rpm the film grows an average thickness of only 0.25 angstrom per revolution.

The low electron bombardment from the magnetron target combined with the 1:9 exposure time cycle result in unusually low substrate temperatures during deposition. Direct temperature measurements are obviously quite difficult, because the temperatures are low and the substrates are continuously revolving. Low temperature is beneficial for the substrate thermal stress, but it is an obstacle for complete chemical reaction during deposit. Therefore addition of radiant heat on the substrates

TABLE II.

COMPARISON OF DIELECTRIC FILMS
PREPARED WITH DC-MAGNETRON
REACTIVE SPUTTERING

FILM	GAS REACTIVITY	PROCESS TEMPERATURE	DIELECTRIC PROPERTIES
SiO_2	High	High	Good
Si_3N_4	Low	Low	Excellent
AL_2O_3	High	High	Medium
GeO_2	High	Medium	Low *

* Water Soluble

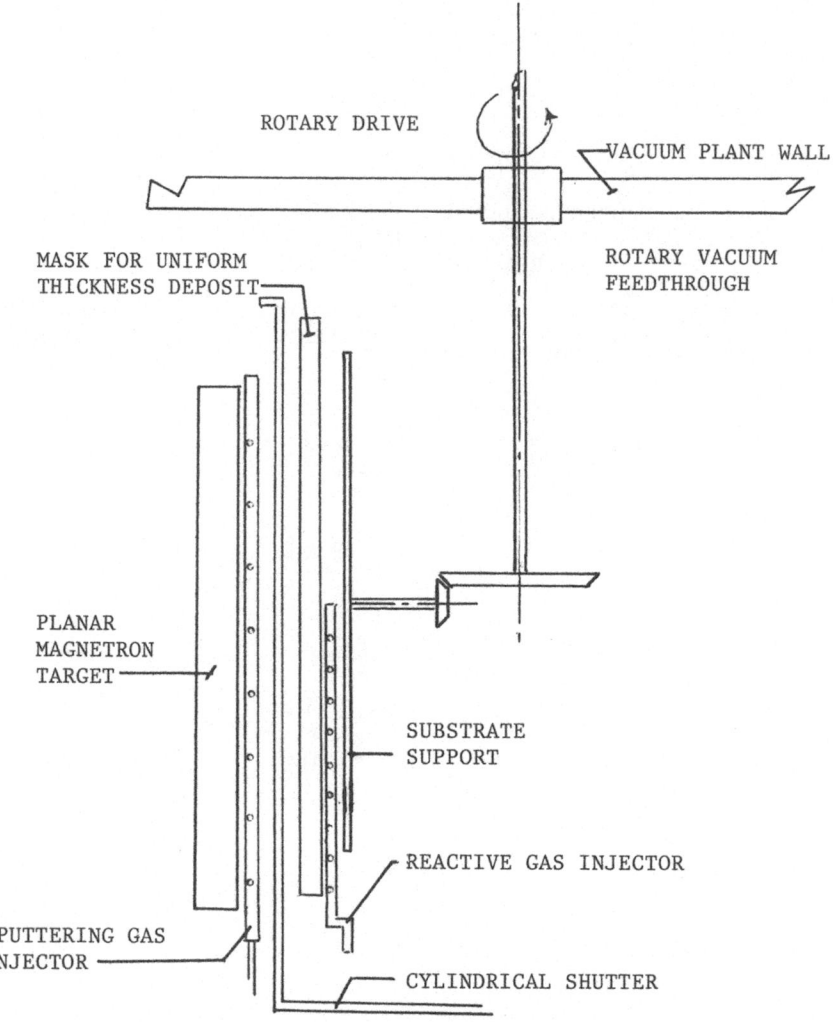

ROTARY DRIVE

VACUUM PLANT WALL

MASK FOR UNIFORM
THICKNESS DEPOSIT

ROTARY VACUUM
FEEDTHROUGH

PLANAR
MAGNETRON
TARGET

SUBSTRATE
SUPPORT

REACTIVE GAS INJECTOR

SPUTTERING GAS
INJECTOR

CYLINDRICAL SHUTTER

Figure 1: CONCEPTUAL LAYOUT FOR THE
PLANAR MAGNETRON REACTIVE
SPUTTERING APPARATUS

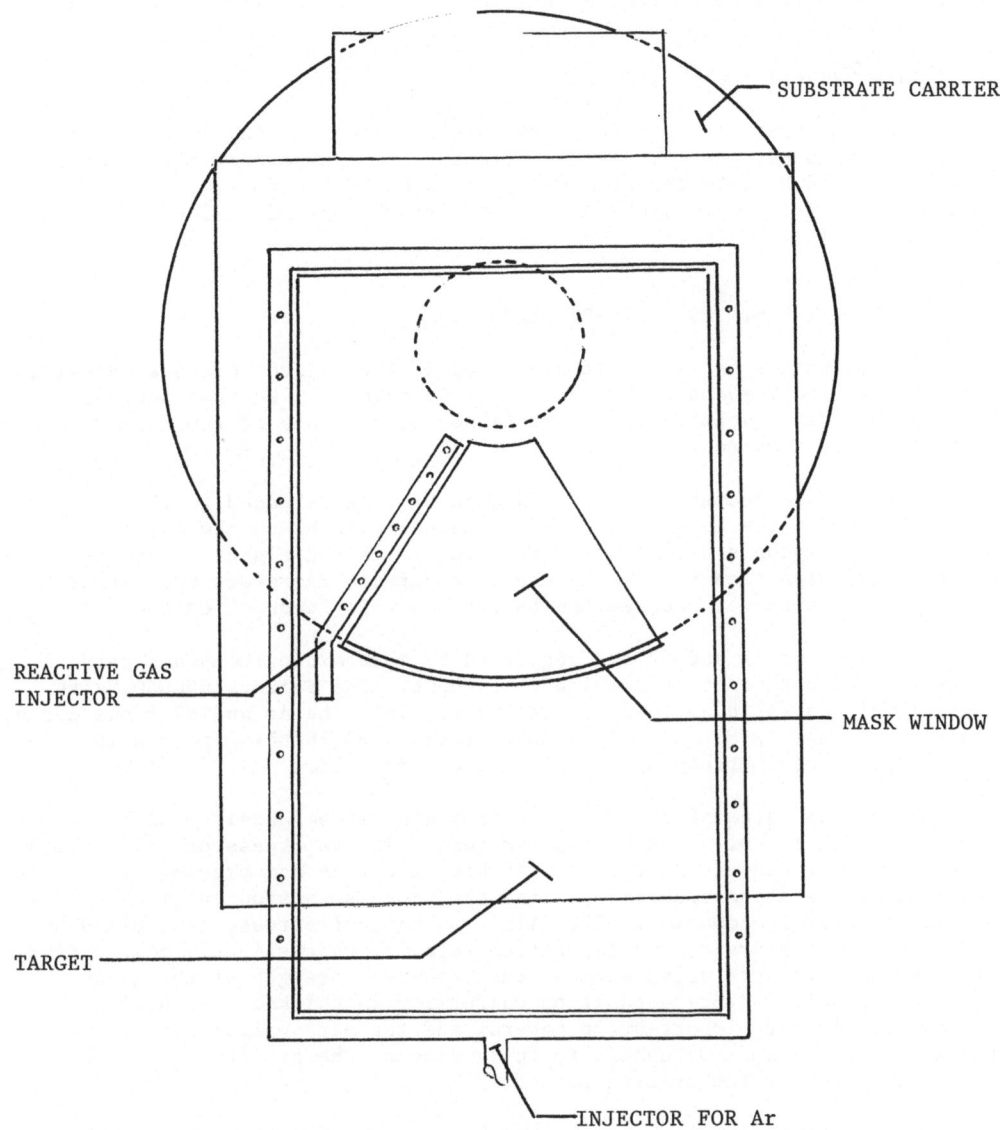

SUBSTRATE CARRIER

REACTIVE GAS
INJECTOR

MASK WINDOW

TARGET

INJECTOR FOR Ar

Figure 2.   PLAN VIEW OF THE REACTIVE
            SPUTTERING APPARATUS

might be at times a necessary compromise for obtaining improved film properties.

The reactive gas injector tube is attached to the mask near the edge of the window, as shown in figure 2. This injector consists of a tube sealed at one end, with perforation ports distributed at intervals, Δ_{xi}, linearly decreasing in length. The length of any of the intervals is related to the next by the following expression, in which k is the linear decrement coefficient,

$$\Delta_{xi} = \Delta_{x(i-1)} - k \, \Delta_{xi}$$

Recalling that the injector ports are situated 1/8 in. in front of the revolving substrates, it seems reasonable to assume that the reaction of the injected N2 with the condensing atoms of Si begins almost immediately after injection, while the concentration of N2 is locally higher than elsewhere.

EVALUATION OF THE SILICON NITRIDE FILMS

Representative process variables used in the silicon nitride deposition are: target power about 1/2Kw and target voltage only several hundred volts. Low target voltage is, as mentioned before, one of the magnetron characteristic features.

This voltage decreases about 10% when N2 flow is added to the Ar flow, because of the formation of Si compounds with N2 on the target. Since these compounds have lower sputtering rates and higher secondary electron emission than Si, as the electron current increases the target voltage must decrease, because the target power is feedback controlled.

The flows of Ar and N2 are regulated by maintaining constant partial pressures against constant pumping throughput. The Balzers RVG040 Valve Control Unit can be used to (indirectly) regulate the Ar and N2 flows during deposition. The Ar/N2 partial pressure ratio used in this process in around 3/2. Total pressures of 1/2-to-1 Pa are used.

Deposition rates of 15 to 25 angstrom/min can be obtained with the appropriate choice of deposition parameters. The thickness of the silicon nitride deposit can be controlled with both accuracy and precision, by controlling the deposition time, since the process is slow and the deposition rate is constant. The film's mechanical stress, controlled by the working gas pressure and deposition rate, is kept slightly compressive. Its adhesion strength often exceeds the cohesive strength of the glass substrate itself. Since deposition uniformity in thickness as well as composition is very important in general and for our application in particular, it seems worthwhile to fully discuss the properties of the films deposited by the present method.

Our apparatus was designed to provide thickness uniformity of the film across the substrates, but careful measurements reveal, as mentioned earlier, some loss of uniformity near the mask boundaries. The analysis was done by probing the film thickness along the radial axis of the disc in nine areas equally spaced over the span of the mask window. The results of these measurements are presented in figure 3. Data from the deposition of a silicon film are plotted on a per cent scale, next to similar data from the deposition of silicon nitride. From this figure it is clear that the film thickness is uniform in the radial direction within 8% over most of the area, but falls off rather steeply in the vicinity of the hub of the disc and near the opposite end of the mask.

This end effect as mentioned earlier is probably due to the reduced target viewing angle from those boundary areas. A reduction of the distance between mask and disc would alleviate the problem, but this would require a costly redesign of the entire apparatus for reducing the mechanical tolerances.

In the course of our thickness calibration experiments, while sputtering with pure Ar, we noticed that the peak thickness of the film (as measured with the technique illustrated in figure 3 shifts in radial position when the angular velocity of the disc is changed. More precisely, the peak "travels" toward the outer rim of the disc when its angular velocity increases. Intuitively it appears that the condensing atoms arriving on the substrate receive a significant radial momentum increment, caused by the centrifugal force. We did not investigate this interesting effect to the full extent, but the radial position of the deposit peak in the mask window expediently dictated our choice of the substrates angular velocity.

The deposition of the reacted silicon nitride film closely follows the thickness distribution of the unreacted Si film (as figure 3 shows). However, the deposit of Si peaks at a slightly different radial coordinate than the silicon nitride deposit. Furthermore, near the inner and outer ends of the window, where the thickness uniformity breaks down, the material reaction appears to be less complete than elsewhere. In fact, the ellipsometric measurements become very difficult, because the film absorption is excessive in those areas.

The refractive index, thickness and dielectric strength have been measured for the films produced by this method and the results are summarized in table III. The refractive index rises from both ends to a maximum value located approximately in the middle of the radial span. The total index variation however is less than 4%, much smaller than the 8% thickness variation measured across the same span. This seems to imply that the chemical reaction proceeds rather uniformly over the span of the window as planned, except again in the boundary regions. The location of the maximum refractive index almost coincides with the location of maximum film thickness. Clearly this is an area of optimum material deposition and reactive gas distribution.

Auger and infrared absorption analysis have determined that the film obtained with this process is deficient in nitrogen with respect to the stoichiometric Si_3N_4. The average refractive index, 1.85, determined by ellipsometry, is considerably lower than 2.00, the bulk Si_3N_4 value. This discrepancy implies that the density and/or the stoichiometry of the reactively sputtered film are substantially different from the bulk value.

Considering however the low deposition temperature, some loss of density and incomplete Si-N reaction are probably unavoidable. During the process development it was observed that when changing the Ar/N_2 ratio during deposition the index increases readily to values near 2.00, but this causes some loss in deposition rates. More complete and accurate data are not available at this time. However, the study of this promising material is still in progress, with the hope to improve understanding of this film's chemical and physical properties and with the aim of controlling surface charge instability and ion drift for further microelectronics applications.

CONCLUSIONS

A method has been developed for the film deposition of silicon nitride by dc-magnetron reactive sputtering, which can produce on a routine basis

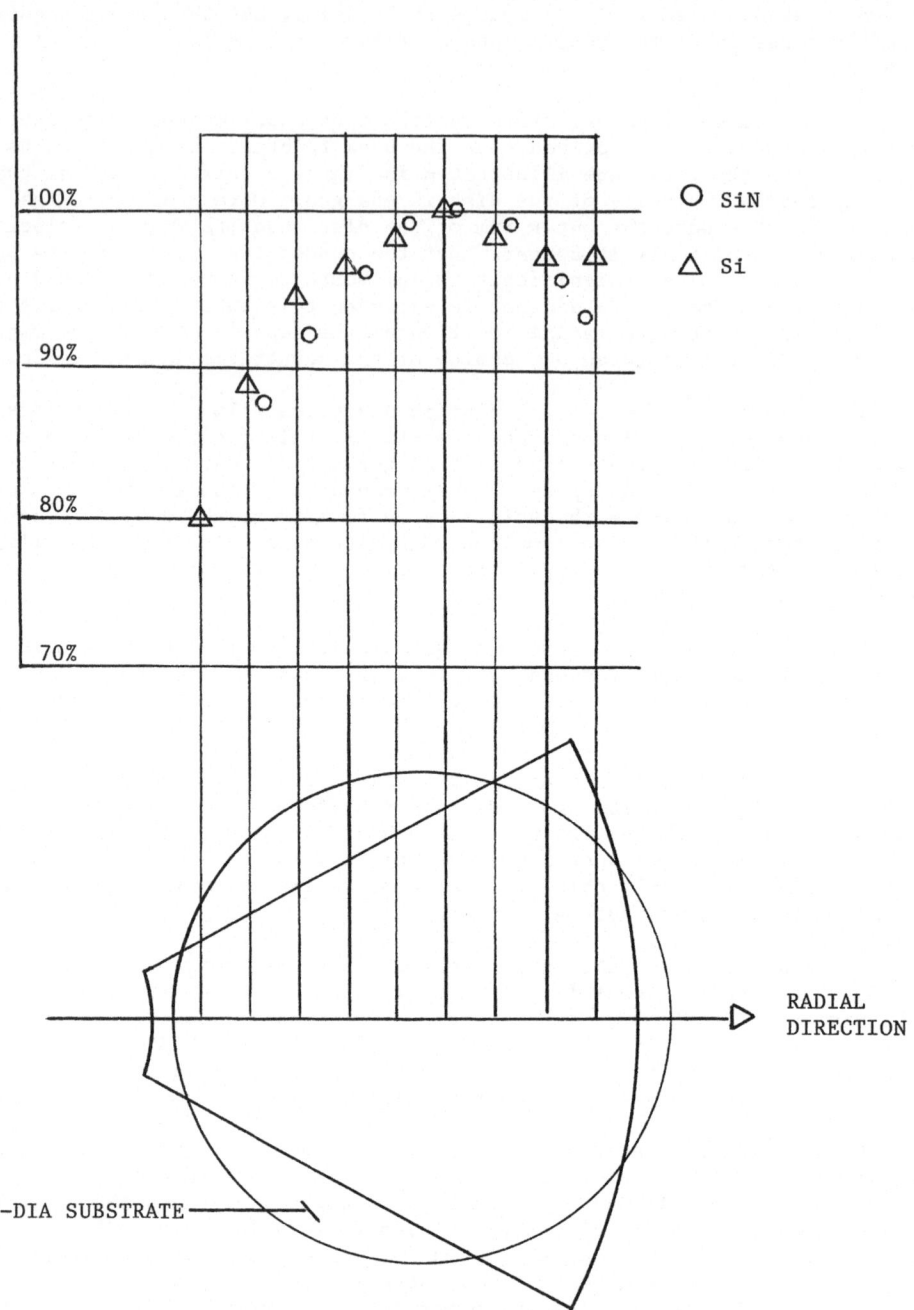

Figure 3. RADIAL THICKNESS DISTRIBUTION FOR
SiN & Si FILMS

TABLE III

UNIFORMITY OF FILM PROPERTIES FOR SILICON NITRIDE DEPOSITED BY REACTIVE DC-MAGNETRON SPUTTERING ON REVOLVING SUBSTRATES

RADIAL LOCATION (IN)	FILM THICKNESS (Å)	N_F	K_F
* 0.3 (1)	–	–	–
0.6 (2)	2289	1.847	0
0.9 (3)	2403	1.863	0
1.2 (4)	2488	1.871	0
1.5 (5)	2560	1.863	0
1.8 (6)	2589	1.857	0
2.1 (7)	2563	1.856	0
2.4 (8)	2464	1.832	0
2.5 (8.1)	2420	1.802	–.001
* 2.7 (9)	–	–	–

1
2
3
4
5
6
7
8
9

SUBSTRATE TEMPERATURE DURING DEPOSITION: 50°C

FILM COMPOSITION BY AUGER ELECTRON SPECTROSCOPY: Si_1N_1

(*) Strong light absorption in the film causes difficult ellipsometric measurement

films meeting all the requirements of mechanical protection, hermetic seal and electrical passivation for many electron devices.

The deposition rate is slow, in order to control the film properties, but the total deposited area is about 28 in. sq. and could be easily scaled up. The film's mechanical adhesion and strength generally exceed the substrate's cohesive strength. The dielectric strength is of the order of several million V/cm. The electrical leakage has not been measured with great accuracy, but appears adequate for our application. Surface charge stability and ion drift instead do not appear yet suitable for capacitor fabrication. However, the low deposition temperature coupled with the absence of aggressive chemical reactants make this process uniquely compatible with complex and delicate thin film structures, including amorphous chacogenide glasses. Uniformity of deposited thickness, film composition and physical properties are still being improved with the steady introduction of refinements in the process variables.

REFERENCES

(1) S. R. Ovshinsky, U. S. Patent 3,530,441.
     Sept. 22, 1970
(2) S. R. Ovshinsky, Phys. Rev. Letters Vol. 21,
     No. 20, 1450 (C)
(3) K. I. Kirov et Al. Thin Solid Film
     41, (1977) L21-L23
(4) D. C. Bartle et Al. Vacuum 33,7,407-410
     (1983)
(5) W. A. P. Claassen et Al. J. Electrochem. Soc. Solid State
     Science, 132,4,893-98 (1985)

OPTICAL FIBERS FOR INFRARED FROM VITREOUS Ge-Sn-Se

I. Haruvi, J. Dror, D. Mendleovic and N. Croitoru

Faculty of Engineering, Dept. of Electron Devices
Tel-Aviv Univeristy
Ramat-Aviv 69978
Israel

INTRODUCTION

CO_2 laser radiation (λ = 10.6 μm) plays a very important role in medicine, communication and material processing. The main difficulty in its successful application in medicine and communication, is the lack of good fibers with low attenuation, thermal and chemical stability, good mechanical properties, low toxicity and ease of fabrication. Several potential materials have been investigated. Polycrystalline fibers [1,3] which have been fabricated by extrusion from materials such as KRS-5, KC1, KBr and the silver halides. Other halides like CsI have been used in growing continuous single crystal fibers [4,5]. From a practical point of view glasses rather than crystalline materials are more desirable for long optical fibers. Low toxic flouride glasses can be used in the near infrared region [6], while chalcogenide glasses are the most promising materials for a wider range of wavelength, near and mid-infrared [7-10]. Although much research on chalcogenide glasses has been done, only limited results are available concerning fiber fabrication and their characteristics [10-14].

The influence of the replacement of Ge atoms with Sn atoms in vitreous Ge-Se system is examined, and the possiblity of using the Ge-Sn-Se chalcogenide glass system for IR fiber production.

The glass formation tendency as well as the structural unit and the character of the immediate environment of Tin in Ge-Sn-Se system, have already been investigated [15-17], using Raman and Mössbauer spectroscopy. According to [15] up to 13a/o Sn might be incorporated into vitreous compositions of the Ge-Se system. The glass formation regions (in the phase diagram) move towards higher Se concentrations (75-80a/o Se). Nuclear Gamma Resonance (NGR) spectra for Ge-Se-Sn alloys indicates that only tetravalent Tin ($SnSe_2$) could be found in the glassy composition. Tin combines only with Selenium and not with Germanium. Ternary compounds of Tin with Selenium and Germanium with intrinsic crystal structure do not exist. However, Tin atoms are incorporated into $GeSe_2$ glassy clusters forming molecular cluster network [16]. For special proportions of the constituents $Ge_{0.65}Sn_{0.35}Se_2$ the molecular cluster network disappears and a continuous network appears [17]. This morphological transition can be understood in terms of reformation of molecular-cluster surfaces in the

247

hetrogeneous phase to yield a homogeneous phase. Although Ge and Sn have the same value for electronegativity (1.8) the Sn-Se bond has a larger ionicity [16] than the Sn-Ge bond and thus, does not form a glass state.

This paper presents some physical properties of the system Ge-Sn-Se. Knowing the intrinsic structure of the glass, we are able to examine how it affects some physical properties (transmission, attenuation , density, micro-hardness, and Tg) of the glasses. We also present a glass fiber drawing experiment.

EXPERIMENTAL

1. Glass Preparation

Ge(6N), Se(5N) alfa chemicals, and Sn(3N) Merck were weighed under inert atmosphere (N_2) in a glove-box. A clean glass ampoule was evacuated and heated (900°C) under vacuum and filled with 8-12 gr of the metal's mixture. The filled tube was then evacuated to 10E-5 torr and sealed. The tube was heated to 900°C at a rate of 4°C/min and was held at that temperature for 20 hours; air quenched and then analyzed.

2. Fiber drawing

Fibers were drawn from sample 2, (Table 1) under inert atmosphere (N_2) in a glove box by the rod method [11]. The drawing system is shown in fig.1.

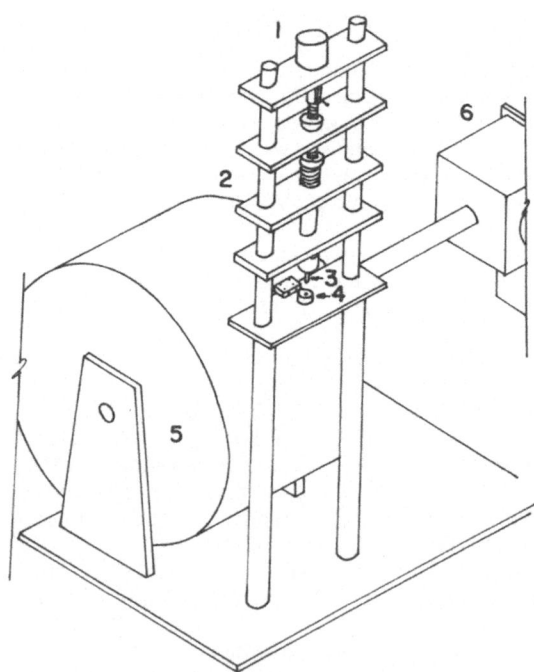

Fig 1.   Fiber drawing system: (1) dc motor (2) clamp (3) glass rod (4) heating coil (5) collecting drum (6) dc motor.

248

A glass rod 6 mm diameter, 20 mm long is mounted in a clamp, that can be lowered or raised by the upper dc motor towards the fixed heating coil. The lowering rate and the heating can be controlled. The first drop of glass is glued to the collecting drum which pulled the heated rod into a fiber.

3. Measurements

a) Composition analysis was carried out with EDAX attached to SEM as can be seen for example in fig. 2.

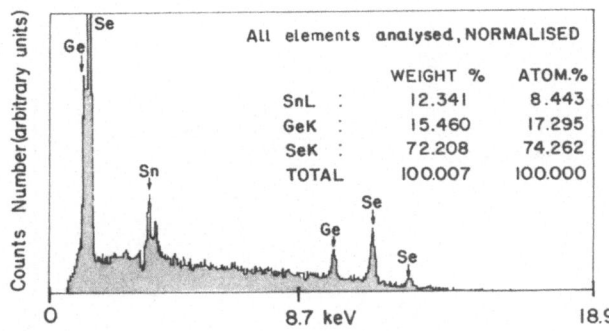

Fig. 2. EDAX analysis of sample 3 (Table 1).

b) X-ray diffraction analyses were carried out on each sample. The lack of any sharp peaks have shown the existence of a glass state.

c) Glass transition (Tg) -DSC (differential scanning calorimetry) thermograms were taken on a METTLER 30 instrument. The 5-10 mg powdered samples were heated under inert atmosphere (N2) at 20°C/min up to 400°C due to safety reasons (Se might evaporate at higher temperatures under experimental conditions). An example of glass transition measurement is shown in fig. 3.

d) The hardness was determined by the Vickers diamond indentation method upon a polished glass surface. The average results of 3 indentations under a load of 100 g on each sample were taken as the final hardness number.

e) Density measurements were carried out by accurate measurement of weight and volume of a sample using freshly boiled water at room temperature (20°C).

f) The transmittance of the glasses was measured with FT-IR (Nicolet-5DX) spectrophotometer in the range 2500-25000 nm, using cut and polished samples.

g) The optical losses through the glass fibers were measured by the cut- back technique, in which the power transmitted by a fiber is recorded before and after a given length of the fiber is removed. The attenuation in the fiber is α_g (dB/cm) calculated by:

$$\alpha_g = 10 \log (p2/p1) / \Delta l \tag{1}$$

where Δl is the length of fiber, p1 the power transmitted before removal, and p2 the power transmitted after removal. A CO_2 laser was used as the light source for 10.6 µm wavelength, while a pyroelectric IR detector (Eltec Model 408) was used for measuring the intensity of the light immediately behind the fiber.

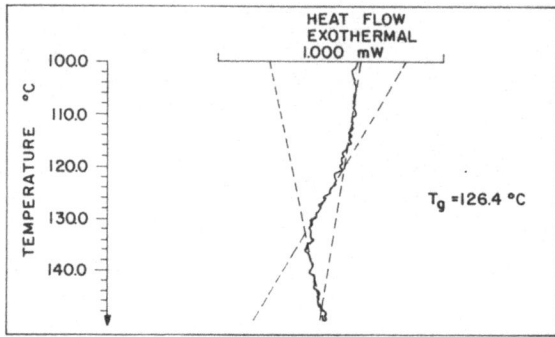

Fig. 3. Glass transition thermogram of sample 1 (Table 1).

RESULTS

In previous papers [15,16] several compositions of the Ge‑Sn‑Se system were obtained. The phase diagram of the system is shown in fig. 4, where our experimental points appear together with points from [15,16]. The glass forming region appears in the left corner of the triangle and is given magnified in fig. 5. As can be seen in fig. 5, there exists some extention in the glass forming region by quenching the melt in ice water.

Our sample concentrations were introduced in the concentration triangle given in [15]. We did not succeed to get glasses with the composition of the form $Ge_{1-x}Sn_xSe_2$, even if the samples were quenched in ice water [16]. As was shown in [17], this could be due to the melt size, which in our case, was about 10 gr. It seems that amorphous samples in the above form can be achieved only for melt size samples smaller than 0.4 gr.

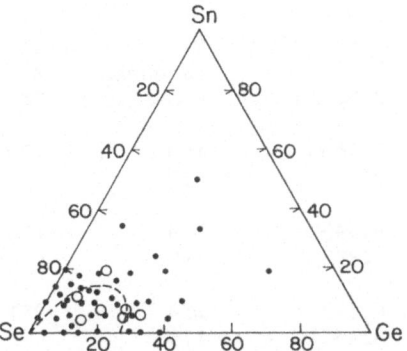

Fig. 4. Phase diagram of the Ge-Sn-Se system. Our experimental points appear as (0), points from [15,16] appear as (.), the dash line is the boundary of the glass formation region.

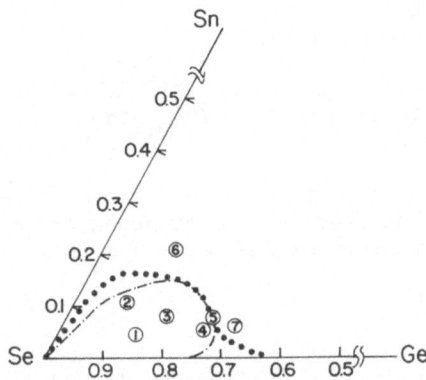

Fig. 5. The amorphous region of the Ge-Sn-Se system. The boundary of the air quenched glass samples is represented as (-.-.-.). The boundary of the ice quenched samples is represented as (......).

Mechanical and thermal properties

In Table 1 the glass samples of the compositions $Ge_{1-x}Sn_xSe_y$, and the values of the density, microhardness, and Tg obtained for them, are listed. The corresponding values for $GeSe_y$, (x=0) are also shown in the table given in [18].

TABLE 1

Composition	Density* d,(g/cm³)	Micro-hardness H,(kg/mm²)	Glass Transition Tg.(°C)	X-ray Diffraction
1) $Ge_{0.76}Sn_{0.24}Se_5$ $GeSe_5$	4.365 4.369	100 113	126 140	amorphous
2) $Ge_{0.4}Sn_{0.6}Se_4$ $GeSe_4$	4.545 4.372	129 147	170	amorphous
3) $Ge_{0.67}Sn_{0.33}Se_3$ $GeSe_3$	4.343 4.355	154 175	211	amorphous
4) $Ge_{0.81}Sn_{0.19}Se_{2.3}$ $GeSe_{2.3}$	4.442 4.34	128	318	amorphous**
5) $Ge_{0.77}Sn_{0.23}Se_{2.3}$ $GeSe_{2.3}$	5.272 4.34		320	crystalline
6) $Ge_{0.37}Sn_{0.63}Se_2$	4.537	21		crystalline
7) $Ge_{0.73}Sn_{0.17}Se_{1.85}$	3.496		355	crystalline

*density is given at 20°C
** possible mixed with crystalline

An example of the transmission (T%) as a function of wavelength (μm) is shown in fig. 6 for sample 2 ($Ge_{0.4}Sn_{0.6}Se_4$) and for a sample of $GeSe_4$ [12] for camparison.

Sample 2 (Table 1) was drawn into a fiber. The attenuation of the glass fiber has been measured by the cut-back method. A value of 1.08 dB/cm for its attenuation at 10.6 μm was obtained.

DISCUSSION

As was shown in table 1 the density (d) of the amorphous samples $Ge_{1-x}Sn_xSe_y$ is practically independent on x. For sample 5 (x=0.23) there appears a sharp increase of the density since the sample is partially crystallized. Unlike the density, the microhardness (H) and glass transition temperature (Tg) decrease when Tin is introduced. This decrease of H and Tg with x can be explained [15] by the assumption of isomorphous replacement of Ge atoms in the structural network of the glass by Sn atoms. The tetravalent Tin, which combines with Selenium, is similar to that in crystalline $SnSe_2$. The $SnSe_2$ has the same crystal structure as CdI_2 (structural unit $^2[SnSe_{6/3}]$), while $GeSe_2$ has a distorted CdI_2 structure (structural unit $^6[SnSe_{4/2}]$); therefore, the formation of hetrostructure solid solutions leads to a considerable weakening of the structure of germanium selenides due to the introduction of Tin atoms.

The comparision between the transmittance of $Ge_{0.4}Sn_{0.6}Se_4$ and $GeSe_4$ (fig. 6), shows that a shift appears towards the longer wavelengths for $Ge_{0.4}Sn_{0.6}Se_4$, starting from the GeO absorption band (~ 13µm) and longer wavelengths (up to 20 µm). This shift is very important for fiber utilization, since it shows that Sn has decreased the contribution of multiphonon absorption to the attenuation.

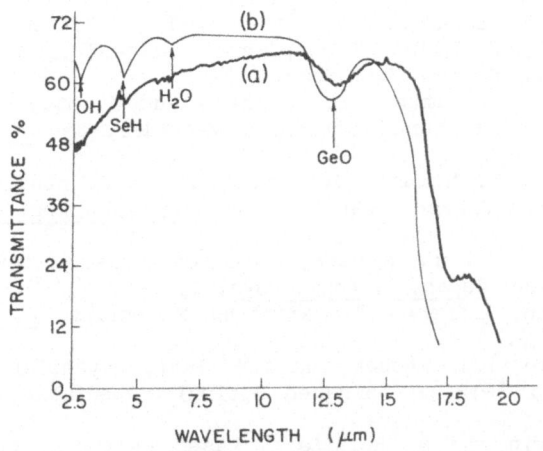

Fig. 6. Curve (a): Transmission spectra of $Ge_{0.4}Sn_{0.6}Se_4$ disc: 10 mm diameter, 1.6 mm thickness. Curve (b): Transmission spectra of $GeSe_4$ disc: 10 mm diameter, 5 mm thickness [12].

Fibers were drawn from sample 2 (Table 1), since for this sample the Tg was approximately the average value between the lowest (126°C) and highest (211°C). The composition of this sample is equivalent to that of $GeSe_4$ since the ratio Se:(Ge+Sn) is equal to the ratio Se:Ge.

The attenuation of our first drawn fibers was lower at long wavelengths (λ >13 µm) than those of fibers drawn from $GeSe_4$. For instance at λ = 16.5 µm an attenuation of 6 dB/cm at fibers drawn from sample 2 and 10 dB/cm at fibers drawn from $GeSe_4$ [12] was achieved.

CONCLUSIONS

1. Samples of $Ge_{1-x}Sn_xSe_y$ were prepared and the conditions of obtaining the vitreous state found.
2. Optical transmission measurements have shown a decrease of attenuation by addition of Sn instead of Ge.
3. Glass transition temperature and microhardness were measured and a decrease of these parameters with the increase of Sn were observed.
4. Fibers were drawn and the attenuation measured.

REFERENCES

[1] A.L. Gentile et al., in: "Optical Properties of Highly Transparent Solids", S.S. Mitra and B. Bendow, ed., Plenum, New York, (1975).

[2] D.A. Pinnow, A.L. Gentile, A.G. Standlee, and A.J. Timper, Polycrystalline Fiber Optical Waveguides for Infrared Transmission, Appl. Phys. Lett. 33:28 (1978).

[3] D. Chen, R. Skogman, E. Bernal and C. Butterin, in: "Optical Properties of Highly Transparent Solides", S.S. Mitra and B. Bendow, ed., Plenum, New York, (1975).

[4] Y. Mimura, Y. Okamura, Y. Komazawa and C. Ota, Growth of Fiber Crystals for Infrared Optical Waveguides, Japan. J. Appl. Phys. 19:L269 (1980).

[5] Y. Okamura, Y. Mimura, Y. Komazawa and C. Ota, CsI Crystalline Fiber for Infrared Transmission, Japan.J. Appl. Phys. 19:L649 (1980).

[6] S. Mitachi and T. Manabe, Flouride Glass fiber for Infrared Transmission, Japan. J. Appl. Phys. 19:L313 (1980).

[7] A.R. Hilton, Infrared Transmitting Materials, J. Electron. Mat. 2:211 (1973).

[8] J.A. Savage, P.J. Webber and A.M. Pitt, Potential of Ge-As-Se-Te Glasses as 8-12 µm Infrared Optical Materials, IR Phys. 20:313 (1980).

[9] A. Bornstein and R. Reisfield, Laser Emission Cross-section and Threshold Power for Laser Operation at 1077 nm and 1370 nm, J. Non-Crystalline Solids 50:23 (1982).

[10] A. Bornstein, N. Croitoru and E. Marom, Chalcogenide Infrared Glass Fibers, in: Proc. SPIE, Advances in IR fibers, 320:402 (1982).

[11] A. Bornstein, N. Croitoru, and E. Marom, Chalcogenide Infrared $As_{2-x}Se_{3+x}$ Glass Fibers, J. Non-Crystalline Solids 74:57 (1985).

[12] T. Katsuyama, K. Ishida, S. Sath, and H. Matsumura, Low Loss Ge-Se Chalcogenide Glass Optical Fibers, Appl. Phys. Lett. 45:925 (1984).

[13] P. Klocek, M. Roth, and D. Rock, The Development and Applications of Chalcogenide Infrared Optical Fibers, in: Proc. SPIE, Infrared Technology XI, 572 (1985).

[14] T. Kanamori, Y. Terunuma, S. Takahashi, and T. Miyashita, Chalcogenide Glass Fibers for Mid-Infrared Transmission, J. Lightwave Technology, LT-2:607 (1984).

[15] P.P. Sergin, L.N. Vasil'ev, and Z.U. Borisova, Mössbauer Effect in Semiconductive Glasses of the System Ge-Se-Sn, Izv. Akad. Nauk SSR, Neorg. Mater, 8:567 (1972).

[16] T. Fukunaga, Y. Tanaka, and K. Murase, Glass Formation and Vibrational Properties in the (Ge,Sn)-Se System, Solid State Commun., 42:513 (1982).

[17] M. Stevens, and P. Boolchand, Universal Structural Phase Transition in Network Glasses, Phys. Rev.B, 31:981 (1985).

[18] Z.U. Borisova, "Glassy Semiconductors", Plenum, New-York and London (1981).

AUTHOR INDEX

SUBJECT INDEX